SpringerBriefs in Environmental Science

SpringerBriefs in Environmental Science present concise summaries of cutting-edge research and practical applications across a wide spectrum of environmental fields, with fast turnaround time to publication. Featuring compact volumes of 50 to 125 pages, the series covers a range of content from professional to academic. Monographs of new material are considered for the SpringerBriefs in Environmental Science series.

Typical topics might include: a timely report of state-of-the-art analytical techniques, a bridge between new research results, as published in journal articles and a contextual literature review, a snapshot of a hot or emerging topic, an in-depth case study or technical example, a presentation of core concepts that students must understand in order to make independent contributions, best practices or protocols to be followed, a series of short case studies/debates highlighting a specific angle.

SpringerBriefs in Environmental Science allow authors to present their ideas and readers to absorb them with minimal time investment. Both solicited and unsolicited manuscripts are considered for publication.

More information about this series at http://www.springer.com/series/8868

K. V. Raju • V. R. Hegde • Satish A. Hegde

Geospatial Technologies for Agriculture

Case Studies from India

 Springer

K. V. Raju
International Crops Research Institute
for the Semi Arid Tropics
Hyderabad, India

V. R. Hegde
Pixel Softek Pvt. Ltd
Bangalore, India

Satish A. Hegde
Pixel Softek Pvt. Ltd
Bangalore, India

ISSN 2191-5547   ISSN 2191-5555   (electronic)
SpringerBriefs in Environmental Science
ISBN 978-3-319-96645-8   ISBN 978-3-319-96646-5   (eBook)
https://doi.org/10.1007/978-3-319-96646-5

Library of Congress Control Number: 2018951781

This Springer imprint is published by the registered company Springer Nature Switzerland AG
The registered company address is: Gewerbestrasse 11, 6330 Cham, Switzerland

Foreword

A great challenge in a country like India, where 142 million hectares of arable land are cultivated by 137 million farm holders, is a realistic assessment of seasonal land use patterns on a real-time basis. This is a crucial missing link in agricultural development in countries like India, as real-time data determines the planning of resources as well as the estimation of food production and associated infrastructure needs. I am pleased to note that this particular study to assess the use of geospatial technology in agriculture was undertaken as a pilot by the ICRISAT Development Center (IDC), ICRISAT, through a project supported by the Standing Panel on Impact Assessment (SPIA) of the CGIAR Independent Science and Partnership Council (ISPC). The study covered three different agro-ecological zones and used remote sensing (RS) technologies together with global positioning system (GPS), geographic information system (GIS), and ground truthing to estimate real-time crop area and status and different land use patterns during a given season. Results from the pilot states have clearly demonstrated the value of using these technologies in agriculture to not only benefit planners in the country but also to be successfully used for risk management initiatives such as crop insurance and in quick settling of claims. This was facilitated through public-private partnership. Smartphones were used to collect three crop seasons' data at farm plot level through geotagging, verification with remote sensing images, and data recorded by the administrative machinery. I am sure this pilot study opens up new opportunities for stakeholders involved in agriculture to use geospatial technologies for efficient decision-making and will also benefit farmers through timely payment of insurance. This book serves as a useful source of information for researchers, students, policy makers, development agencies, and corporations involved in real-time planning and management in agriculture.

Research Program Director, Asia and Director,   Suhas P. Wani
ICRISAT Development Center
Hyderabad, India

Foreword

 A lack of real-time data that captures information on farmers' land holding patterns and season-wise cropping patterns is a significant deficiency for agricultural decision-making at a range of scales. Lately, ICRISAT has initiated several activities in the sphere of digital agriculture. In the three Indian states of Odisha, Andhra Pradesh, and Karnataka, representing different agro-ecosystems, land use data were collected through satellite imagery, ground truthing, and government records as a pilot to develop satellite-based land use planning data at the village level. A representative village was identified in each of the states in consultation with the National Remote Sensing Center and the respective state governments.

A meticulous job was done of capturing and analyzing these data, which is highly appreciated by all the three state governments. This first-of-its-kind exercise was supported and guided by the Standing Panel on Impact Assessment (SPIA) of the CGIAR Independent Science and Partnership Council (ISPC). I am thankful to all who supported this activity.

I am sure this study will provide useful insights to researchers, students, and policy makers in understanding the micro-level situation at the village level and the abundant opportunities for scaling-up the use of low-cost and time-saving digital technologies.

Deputy Director General - Research   Peter Carberry
ICRISAT
Hyderabad, India

Foreword

One of the major missing links in agriculture is capturing farmers' database laced with their land holding pattern and season-wise crop pattern on digital platform. In ICRISAT, in recent years, we have initiated several activities in this regard. As part of this process, we had initiated capturing the ground truth in three different states of India—Odisha, Andhra Pradesh, and Karnataka. In each of these states, in consultation with the National Remote Sensing Center and the respective state governments, a representative village was identified.

In these village-level, micro-level studies, the crop pattern data on a digital platform using mobile phones captured three crop seasons data from the farm plot level. Our team of authors have done meticulous work in capturing and analyzing this work. This was highly appreciated by all three state governments. The first of its kind exercise was supported and guided by the SPIA of CGIAR group. I am extremely thankful to them for supporting this activity.

I am sure this study will be useful in understanding the micro-level situation and opportunities to scale up by using digital technology at low cost and in lesser time. Surely, this study will be useful to all the researchers, students, and policy makers in the years to come.

Director General, ICRISAT   David Bergvinson
Hyderabad, India

Preface

One of the core objectives of the Sustainable Development Goals is doubling the farmers' income. Reliable and real-time information related to agriculture is essential to address agricultural growth in developing nations. Agriculture data includes agricultural holding by distribution, size, tenure, land use, means of production, and labor force. The links between poverty and crop yields that depend upon factors like cultivation practices, availability of irrigation, and access to resources to buy agricultural inputs for adoption of new technology can be understood only with reliable information of crop area and types. The quality of available agricultural data and the methods by which such data is collected are weak in several developing countries. As of now crop area estimates have problems associated with both accuracy and timely availability. Crop area data are generally available a few months after harvest, and having reliable data before harvest is a major challenge. There are useful methods that generate information at the national and state level, while at the village level there is a need for spatialized information about agricultural practices.

Toward addressing the gap in reliable agriculture statistics at the village level, ICRISAT conducted action research in three different agro-climatic regions of India by using geospatial tools. The theme of the research has been exploring means of generating reliable information on crops during the growing season. Geo-stamping approach, i.e., remote sensing (RS), global positioning system (GPS), and geographic information system (GIS), has been adopted in conformation with existing work flow for crop area enumeration. Plot level crop inventory using geospatial tools, along with some important socio-economic aspects of the farmers, revealed the gaps in the current approach and provided useful insights into the farming practices and constraints of farmers. The exercise also analyzed kitchen gardens within the settlement, where women are involved in growing vegetables that contribute to household income.

The method adopted can be scaled up as it improves the crop statistics in terms of accuracy, authenticity/reliability, traceability, timeliness, access, and analysis. Since the data is organized as per the administrative hierarchy, crop area information even at the holding level is available and there is no delay in compilation. The methodology is simple and can be devolved to youths in the village as it provides for temporary employment opportunities.

It is hoped that the information brought out in this book will be useful not only to the governments of the three states (Odisha, Andhra Pradesh, and Karnataka), but also to those elsewhere in the world who look forward to generating reliable and accurate crop statistics. Also, the process has inbuilt employment options for youths who could make it a profession.

Hyderabad, India   K. V. Raju
Bangalore, India   V. R. Hegde
Bangalore, India   Satish A. Hegde

Acknowledgments

This book is based on a larger study supported by the Standing Panel on Impact Assessment (SPIA) of CGIAR's Independent Science and Partnership Council (ISPC) mainly under the grant "Strengthening Impact Assessment in the CGIAR (SIAC)." We are extremely thankful to them for their continuous guidance and encouragement.

We are highly thankful to Dr. David Bergvinson, Director General, ICRISAT; Dr. Peter Carberry, Director General (Acting); and Dr. S P Wani, Director, Research Program—Asia for their full support and encouragement to carry out this study in three states of India. Our special thanks to Dr. Kumara Charyulu for being a part of this team in its initial stage and supervising the fieldwork in Odisha and Andhra Pradesh. We highly appreciate his meticulous hard work in this regard.

We gratefully acknowledge the Governments of Andhra Pradesh, Karnataka, and Odisha for providing required coordination during the stages of the experiment. We sincerely thank Shri Shiva Rama Krishna (Mandal Revenue Officer, Gantasala); Shri K.V. Subramanyam (Director, Dept. of Economics and Statistics, Government of Karnataka) for providing useful information and useful discussion on the subject; and Shri Subhas Chandra Biswas (Deputy Director of Agriculture, Angul district, Odisha) for providing necessary cooperation and support in Angul district.

We extend our sincere thanks to P V Rai, MD, and our colleagues Basavaraj Patil, K. Nagaraja, and Jayalakshmi (all from Pixel Softek, Bengaluru) for the help in generating spatial data and validating the mobile application, and organizing the data in GIS.

We are thankful to Ms. Suchita Vithlani for her meticulous administrative support during the entire period of study.

Contents

Chapter 1
Introduction

Abstract World needs to elevate 700 million poor people living in rural area. Sustainable Development Goal emphasizes ending poverty and hunger by 2030 and the importance of agriculture in the economy, employment, food security, national self-reliance, and general well-being. Reliable information related to agriculture is at the center of everything and agricultural statistics are part of the economic profile of a village/taluk/district/state and country and are increasingly becoming important. Agriculture data includes agricultural holding by distribution, size, tenure, land use, means of production, and labor force and statistics are essential to monitor trends and estimate future prospects for agricultural commodity markets which can assist in setting policies such as price supports and strategies to promote economic growth.

Keywords Poverty · Food security · Agricultural statistics · Agricultural holding

In 2010, over 900 million poor people (78% of the poor) lived in rural areas, with about 750 million working in agriculture (63% of the total poor). About 200 million rural poor could migrate to urban areas by 2030, based on urbanization projections and assuming migration of a proportional share of the rural population that is poor (if three of every ten people who migrate are poor). This would leave about 700 million poor people in rural areas to be lifted out of poverty by 2030. Even with a projected increase of the share of the population living in urban areas in less developed regions (increasing from 46% in 2010 to 56% in 2030), population growth is projected to still lead to a small net increase in the number of people in rural areas from 3.06 billion to 3.13 billion (Townsend 2015).

The extract from the World Bank Report (2015) emphasizes the importance of agriculture in the economy, employment, food security, national self-reliance, and general well-being. Ending poverty and hunger by 2030 being one of the core objectives of Sustainable Development Goals (World Bank Report 2015) speaks of doubling the income and the need for improved agricultural productivity and climate resilience and strengthened links to markets, agribusiness growth, and rural non-farm incomes. Therefore, the reliable information related to agriculture is at the

© The Author(s), under exclusive licence to Springer Nature Switzerland AG 2019   1
K. V. Raju et al., *Geospatial Technologies for Agriculture*, SpringerBriefs
in Environmental Science, https://doi.org/10.1007/978-3-319-96646-5_1

center of everything. Agricultural statistics are part of the economic profile of a village/taluk/district/state and country and are increasingly becoming important.

Agriculture data includes agricultural holding by distribution, size, tenure, land use, means of production, and labor force. Agricultural statistics are essential to monitor trends and estimate future prospects for agricultural commodity markets which can assist in setting policies such as price supports and strategies to promote economic growth. In addition to its inevitable role in food security, agricultural development is now seen as a vital and high-impact source of poverty reduction. The links between poverty and crop yields, depending upon a variety of factors like cultivation practices, availability of irrigation, and access to resources to buy agricultural inputs for adoption of new technology, cannot be fully understood without reliable information of crop area and types. In the absence of reliable information on crop productivity, the reasons behind food insecurity of agricultural households cannot be precisely identified.

Though importance of agriculture in economic development and food security is being addressed, the quality of available agricultural data and the methods by which such data is collected are weak in several developing countries. As of now, crop area estimates have problems associated with both accuracy and timely availability. Crop area data are generally available a few months after harvest. Having a reliable data before harvest is a major challenge. Though recent developments in remote sensing and related geospatial tools provide access to new data and methods for strengthening data systems, their application has been limited (Miller 2010). This book addresses the gap by examining the current data systems and demonstrating the significant potential role for geospatial tools in improving the quality of agricultural data and the methods by which it is obtained and thereby adoption of the same in developing countries. The recommendations are based on the experiments conducted by International Crops Research Institute for Semi-Arid Tropics (ICRISAT) in three different agroclimatic regions of India.

The objectives of this book are fourfold. First, a literature review of different methods and the recent technological tools adopted for crop statistics are provided. Second, the process and practices followed in India are documented. Third, the practices followed in three states of India are analyzed. Fourth, the rationale and method deployed for crop inventory along with results are described, and a comparative analysis and recommendation to enhance data quality are presented.

Chapter 2
Crop Area Statistics

Abstract Generation of crop statistics in India dates back to Kautilya's Arthashastra (an ancient Indian treatise on statecraft belonging to third century BC) as well as Moghul era (sixteenth century). Currently, the crop statistics are generated based on land revenue system for major food crops and non-food crops. The data is received from the State Agricultural Statistics Authorities in various states and union territories. Methods for crop area estimation based on different sampling techniques have been successful but cost-effective methods, especially in developing or underdeveloped nations, are needed. New technologies like remote sensing, GPS, and GIS have played a major role. Crop area estimation at the national level is more established. Addressing accuracy first, it is important to address national versus small area estimation; in terms of accuracy, it seems to be a choice between accuracy and cost, assuming each has an approximate level of timeliness. Methods using satellite images have the essential component of reference of ground truths. In a mixed cropping pattern with small and fragmented holdings, the extent of ground truths was found to be inadequate, and studies indicate that manual extraction of field boundaries with thorough knowledge of the landscape provides useful results and provides control on datasets for further validation, while field inventory is done adopting an integrated approach.

Keywords Crop statistics · Land revenue system · Satellite images · Mixed cropping pattern · Fragmented holding · Ground truth · Field boundaries

2.1 Rationale

Area covered by any crop and the yield per unit area are the important components of crop production estimation. Though information on production is the priority, the area estimation is essential since it is subjected to variations over the seasons/years due to abandonment, extreme weather, or unusual economic conditions, while the yield remains relatively the same from a unit area. The Group on Earth Observations

© The Author(s), under exclusive licence to Springer Nature Switzerland AG 2019   3
K. V. Raju et al., *Geospatial Technologies for Agriculture*, SpringerBriefs
in Environmental Science, https://doi.org/10.1007/978-3-319-96646-5_2

(GEOSS 2008) while reviewing the practices of crop mapping using remote sensing highlights the value addition of newer technologies as follows:

*   The inaccuracy of estimations is smaller than the uncertainty of the information previously available.
*   The geographical detail of the estimates provided is finer than the existing information.

The review emphasizes the reliability of the information from the newer methods and also the cost at which information is generated. The importance of geographical context should always be associated with the information as it helps in timely decision related to disasters or managing insurances. While the overall goal is toward doubling the income and improved agricultural productivity and climate resilience, the need for spatialized information about agricultural practices is expected. Information on agriculture practices, crop varieties, crop succession, and cropping pattern techniques, and the infrastructures are to be the part of agriculture statistics. FAO's World Programme for the Census of Agriculture (FAO 2005) recommends that censuses consider the agricultural holding or farm as the basic unit for production and other economic statistics.

Addressing the need of useful technological solutions for poor farmers in less favored environments of Asia and elsewhere, IFPRI advocated (Pender 2007) for the options that are suited to specific locations. This requires a pragmatic approach to learning what works well where and why. In pursuit of such pragmatic options for farmers, research and development programs should not ignore the potentials of traditional farming practices, which are well suited to farmers' needs in many contexts. It indirectly emphasizes the geographic context and constraints at farm level and the real-time, reliable information on agriculture practices.

2.2   Approaches

There have been different approaches to crop area estimation across the world. Crop statistics have been generated by censuses by means of enumeration of the total population of interest and also by samples of only a small part of the larger area. Earlier ones involved expert opinion of local voluntary crop reporters in the villages, though one of the cheapest forms of collection of data was not always accurate (Craig and Atkinson 2013). Total enumeration of all the people and later all the farmers in the villages has been another approach; census information is usually not cost-effective and takes more time. An early paper (Huddleston 1978) which documented sampling and estimation strategies for crop forecasting and estimation is still being considered relevant. Though there are rich literatures related to crop statistics based only on ground surveys, there have been refinements towards collection and analysis of land use and other natural resources.

The Natural Resource Conservation Service (NRCS) of USDA has an extremely large point sample and survey known as the National Resources Inventory (NRI).

The objective of NRI is to monitor "status, conditions and trends in soil, water and other natural resources data on non-Federal lands in the United States" (Breidt and Fuller 1999). The data gathered include broad land use and land cover categories primarily using photo interpretation and by ancillary data sources in house (Nusser and Goebel 1997). The USDA's National Agricultural Statistics Service (NASS) has a conventional area frame-based program which uses natural boundary sample segments with personal interviews of the farmers operating the land in those segments (Davies 2009; Cotter et al. 2010). It is the combination of area- and list-based information through a multiple frame sampling approach that leads to complete and cost-efficient results for agricultural statistics of all sorts.

In most of the countries where ground surveys are conducted, either for standalone estimation or for remote sensing, ground truth data have been an obvious and common choice. Area frame sampling methodology adopted in MARS project (launched with the support of the European Commission and Eurostat) focused on crop area and production estimation of annual crops: soft and durum wheat, barley, rapeseeds, dried pulses, sunflower, maize, cotton, tobacco, sugar beet, potatoes, rice, and soya, as well as fallow on arable land (Gallego 1999). The Southern African Development Community (SADC) countries use a combination of subjective procedures (i.e., extension officer's and/or grower's assessment) and objective ones involving direct measurement. In most of the SADC countries, even those with some of the better crop forecasting procedures in place, crop forecasting has been limited to cereals and other major crops.

Gradually, the Geographic Information Systems (GIS), the technology for data integration, got introduced into the crop statistics. Earlier GIS was mainly used to store final area sampling frames in a digital form, but as the technology increased, they became a part of the construction process. Later, loading satellite images directly in to GIS started with related image processing programs. Being able to review and store boundaries derived from or overlain on different data sources, GIS greatly aided area frame construction. New devices such as Global Positioning System (GPS) receivers and various types of handheld computer tablets, etc. have been deployed for data capturing anywhere with some limitations (Carfagna and Keita 2009).

Slowly with the advent of remote sensing and GIS, integrated approaches for crop surveys started. A case study in South Africa (Ferreira et al. 2006) wherein integration of remote sensing, point frame sampling, GIS, and aerial observations was adopted and this became a system later called Producer Independent Crop Estimates System (PICES). Satellite imagery (Landsat 5) over three seasons was used to capture field boundaries for all cultivated fields. The field boundaries replacing the crop density stratum were used for defining a frame for a random geographic systematic selection of sample points across each province. The crop type for each sampled field/point was determined by aerial observations by trained observers (predominantly farmers) using light aircraft. The crop information was recorded on a digital tablet PC in combination with a GPS navigation instrument. Information gathered during the aerial survey was used both to calculate a statistical area for each crop type per province and as a training set for satellite imagery for classification purposes, the latter use of which resulted in a complete set of classified fields for each province. The method sets a trend for an integrated approach.

2.2.1 Remote Sensing in Crop Area Estimation

Remote sensing techniques have become popular in area estimation over the past few decades, as the technology and methodologies have matured. Since the inception of civilian remote sensing program in the United States in the early 1960s, a major research and development thrust has been on agricultural crop identification and area estimation (Dadhwal et al. 2002). Experiments such as Crop Identification Technology Assessment for Remote Sensing (CITARS) and Large Area Crop Inventory Experiment (LACIE) were conducted to demonstrate the capabilities of remote sensing for crop inventory and forecasting (MacDonald 1984). The experiment not only demonstrated the usefulness of automated data processing techniques and spaceborne data for corn and soya bean inventory in the United States; it also proved operational capability of remote sensing technology for wheat production forecasting (MacDonald and Hall 1980).

USDA-NASS initiated its remote sensing acreage estimation program in the 1970s and early 1980s with the Large Area Crop Inventory Experiment (LACIE) and Agriculture and Resources Inventory Surveys through Aerospace Remote Sensing (AgRISTARS), to determine if crop acreage estimates could be derived using multispectral imagery and ground truth data (Bailey and Boryan 2010). These programs were successful at generating unbiased statistical estimates of crop area at the state and county level and reducing the statistical variance of acreage indications from farmer-reported surveys (Mike 2010). NASS' remote sensing acreage estimation program evolved over the years paving the way for the current GIS-based Cropland Data Layer (CDL) program which has been in existence since 1997.

Approaches adopting remote sensing-based crop inventory and crop discrimination are based on differential spectral response of various crops in a multidimensional feature space produced by different spectral bands, or time domain, or both (Dadhwal et al. 2002), and for plantation crops, the canopy architecture (Hegde et al. 1994) is being referred to. An excellent document on the issues of area estimation in general as well as those from the Earth Observation (EO) perspective was published by the Group on Earth Observations (GEO) following a June 2008 conference on the topic (GEOSS 2008). Area estimation throughout the crop season is typically accomplished through ground surveys or ground surveys supplemented with remotely sensed data. The remote sensing imagery is generally used for stratification and is often used directly in estimation as well. One of the key issues highlighted in this conference was the importance of quality ground surveys. In spite of the technological developments and capabilities of remote sensing, quality ground surveys, where at all possible, are an extremely important (and nearly indispensable) piece of the area estimation process.

In practice, high-resolution imagery with image interpretation is sometimes substituted for ground surveys, but this proxy for a ground survey is generally used only in response to budgetary limitations, access restrictions, or simply the desire for more timeliness (i.e., avoiding the time of collecting and processing the ground data). Most of these efforts have been less than totally successful. For example, Gallego (2006) and Narciso et al. (2008) describe two recent research efforts focused on avoiding or minimizing ground data collection that ultimately proved unsatisfactory.

With technology improvements in sensor quality and availability and processing advances, it has become an even stronger player in countries' efforts to estimate crop area and production. Issues, especially in regard to timing, efficiency, and assurances of continuing availability, remain with its use. According to GEO Ag Task 07 03 (Gallego et al. 2008), the timing or schedule of crop area estimation or early estimation depends on the following elements: (1) the number of days after sowing a crop that can be detected by a remote sensor, (2) the spatial variability in sowing practices of the region, (3) the crop calendars of competing crops, (4) the characteristics of remote sensors (revisiting time), (5) the date in which the crop can be reliably recognized in the field, (6) the time needed for the ground survey, and (7) the time needed for ground data processing.

Crop area estimation and condition assessment (Kussul et al. 2012) in Ukraine, under the auspices of the Joint Experiment of Crop Assessment and Monitoring (JECAM) project of the Global Earth Observation System of Systems (GEOSS), used information from various satellites, including MODIS data from the Terra and Aqua satellites (from the RC Agri4Cast Image Server at http://cidportal.jrc.ec. europa.eu/thematicportals/agri4cast); Thematic Mapper (TM) data from Landsat-5, Earth Observer 1 data provided by the National Aeronautics and Space Agency (NASA); and data from the Ukrainian satellite Sich-2 launched in 2011. As the source for ground truth data, three Ukrainian test sites were set up and are being used.

In India, various experiments were being conducted to improve the crop production forecasting, and (Lochan 2006) the most current one is called Forecasting Agricultural output using Space, Agrometeorology, and Land-based observations (FASAL). The predecessor of this initiative was Crop Acreage and Production Estimates (CAPE), which was first launched in 1987 to utilize remote sensing techniques in crop area and production forecasting. The CAPE project has successfully achieved national-level forecasts of wheat and *Kharif* rice, in addition to making district level pre-harvest production forecasting of cotton, sugarcane, rapeseed/mustard, and *Rabi* sorghum in major growing regions in the country by using remote sensing technology and other auxiliary information. It has overcome the problem of persistent cloud cover during the *Kharif* season by using SAR data from RADARSAT.

Indian experience at crop forecasting using remote sensing technology focusing on wheat production in the state of Haryana (Hooda et al. 2006) provides a time series of how well remote sensing-based estimates have compared over the years for both crop area and production. Improvement in the resolution of sensors and methodology, as well as increased computing power, has clearly improved the estimates over time. Remote sensing imagery is used for both stratification and estimation. The stratification is on intensity of agriculture and is updated every 3–5 years. Segment sizes have decreased from 10×10 m to 7.5×7.5 m and finally to 5×5 m, while sampling rates have increased. Some of the small districts in the state of Haryana are now completely enumerated.

In one of the studies (Jain et al. 2013) of two regions in India (Gujarat and southeastern Madhya Pradesh—that represented diversity in crop type, soils, climatology, irrigation access, cropping intensity, and field size), different methods were tried to quantify cropping intensity of smallholder farms where the size of one field is typically smaller than the spatial resolution of readily available satellite data. In the study, multi-scalar datasets were used to assess cropping intensity of smallholder farms: (1) the Landsat threshold method, which identifies if a Landsat pixel is cropped or uncropped during each growing season; (2) the MODIS peak method, which determines if there is a phenological peak in the MODIS Enhanced Vegetation Index time series during each growing season; (3) the MODIS temporal mixture analysis, which quantifies the subpixel heterogeneity of cropping intensity using phenological MODIS data; and (4) the MODIS hierarchical training method, which quantifies the subpixel heterogeneity of cropping intensity using hierarchical training techniques. The results specifically apply to our study regions in India and most likely also apply to smallholder agriculture in other locations across the globe where the same types of satellite data are readily available.

Various researchers have proposed procedures for crop area estimation in which visual interpretation of high-resolution imagery of a sample of grid points is substituted for ground observations. This particular report trumpets the advantages of this in developing countries where a structured statistical ground survey may be too costly or otherwise infeasible.

A new survey method for area estimation, called "the dot sampling method," was adopted for a rice planted area survey in Sri Lanka and Thailand (Jinguji 2014). This new method has been developed by combining a traditional attribute survey method with two current information technologies, i.e., Excel and Google Earth. These combinations enable to achieve a simpler, reliable, and cost-effective survey method in comparison with existing methods. The approach that was developed to avoid sampling of nonagricultural lands allows selection of a sample point (dot) on the ground identified by its coordinates (latitude, longitude) and overlaid on the satellite imagery like Google Earth of the survey area, leading to the selection of plots/fields. The frequency distribution of sample dots is generated as per land use categories of interest including crop area, which when multiplied by the total survey area gives the estimate of land use under each category as well as its standard error. The method can estimate not only core crops area but also minor crops area in a whole target area during every crop season in a year. It is interesting to note that there are no measurements involved in the method, and it is therefore free from measurement-linked non-sampling errors.

A combination of field server and satellite remote sensing technology for crop monitoring in the Philippines highlights the ability of remote sensing in identifying crops on the field and in detecting changes and classification of agricultural produce

(Labuguen et al. 2014). The study "Agricultural Land Information System (ALIS)" also showed the capability of the technology in distinguishing the single crop and multi-crop planting patterns on the ground. This was possible because the crop signatures (remote sensing signals) are different for these cropping patterns. In addition, in situ data is necessary for calibration and validation of the product derived from satellite images. Since collecting in situ data is costly and time consuming, field server (FS) technology was used to collect field data. FS is an Internet Field Observation Robot that consists of a set of multiple sensors, a Web server, an Internet Protocol (IP) camera, as well as wireless interfaces. It is designed to provide an outdoor solution for environment monitoring. A similar study was also conducted in Thailand using field server and satellite remote sensing for rice crop monitoring (Rakwatin et al. 2014).

Geospatial technologies (remote sensing and Geographic Information System) have been used to assess the agricultural potential of the Nebo Plateau, a rural area in the Limpopo Province of South Africa (Petja et al. 2014). The approach entails assessing the suitability in terms of land/soil and climate, which are determinant factors for agricultural development. Though not specifically focused on crop area enumeration, the study combined different geospatial tools to develop a decision support system.

With the advancement of technology in terms of spectral as well as spatial resolutions, numerous studies have been conducted: (1) crop discrimination at the early stage of growth using multi-temporal high-resolution data (Song et al. 2017), (2) crop area mapping using spectral matching and phenological characteristics at different locations across the geographies (Özüm Durgun et al. 2016), and (3) crop classification based on single images from very high-resolution (VHR) satellite sensors (Ozdarici et al. 2015).

A thorough review and analysis of remote sensing related to agriculture (Bégué et al. 2018) show that the remote sensing community has focused on the detection and characterization of agricultural practices mostly limited to case studies. The authors have also correlated the terminology and common definitions for different agronomic conventions (Fig. 2.1) for the review of the literature. The shortcomings can be explained by the wide variety and variability of agricultural practices, which cannot be properly captured and described at the plot scale over large areas, due to the lack of suitable satellite data, such as dense time series of optical and radar images at decametric resolution. The increasing availability of remote sensing data, in particular the free European Sentinel-1 and 2 constellation data suitable for small to medium field size monitoring, and the emergence of new data processing techniques such as data mining and deep learning should stimulate the research in this area.

The review of various studies and experiments on crop area estimation indicates that the practices and methods are mature and multiple methods (area sampling, list sampling, point sampling with or without a grid) have proven to be successful. One area that probably may need continuous research is for more cost-effective methods, especially in developing or underdeveloped nations. Thus, any research aimed at reducing the cost of sampling frame construction or data collection would seem to

Fig. 2.1 Typology and short definitions of the cropping system components (Bégué et al. 2018)

have a role. New technologies have played a major role with GIS, GPS, remote sensing, etc., but there seems to be new ones every year or so that could be examined. Crop area estimation, at the national level, is more established and mature than crop yield forecasting. Addressing the accuracy first, it is important to address national versus small area estimation; in terms of accuracy, it seems to be a choice between accuracy and cost, assuming each has an approximate level of timeliness.

2.2.2 Studies on Microlevel Mapping: Integrated Approaches

There have been continuous efforts to improve the crop area statistics, and one of the strategies of restructuring recommended is to distinguish between:

- The need for reliable and timely flow of data on area and production on a regular basis for every crop season, year after year
- Requirements for various special purposes such as crop insurance

The reviews have also recommended the expansion of the present remote sensing program and provide reliable and validated in-season forecasts and end-season estimates of area for a wider range of crops at the state and national levels, as well as comprehensive and detailed plot-level data of land use and crops at the village level.

The experiments on using satellite remote sensing with high spatial resolution have been effective, but still the technology appears to have limitations in the regions (as in the case of India—semi arid tropics) with fragmented holding and mixed cropping system. Cloud coverage is one of the factors that hinder the use of remote sensing, and further in a mixed cropping system, it becomes difficult to delineate different crops. Scientists have been trying integrated approaches for mapping on large scale using other geospatial techniques.

Combination of images from multiple satellites and better spatial and spectral resolution are being tried mainly to get cloud-free data and to assess suitability for different crop signatures (Singh et al. 2005). While reviewing the use of LISS-III data for village-level crop mapping that indicated about 90% accuracy of area is possible for crops with larger aerial extent, Vijay et al. (2013) have made an attempt to correlate the village parcels with LISS-III and LISS-IV images and found mis-classification of crops with smaller size of the crop plots. The accuracy was only 70–75% and have brought out the limitations of remote sensing-based village-level crop inventory at plot level. Panda et al. (2010) have reviewed the utility of various remote sensing data from different platforms like satellite, LIDAR, aerial, and field imaging for site-specific crop management which address mapping and GIS-based modeling. Looking at the need of reliable data on crop statistics and their applicability in various sectors, continuous updation of technologies; GIS, remote sensing, and mobile GPS has been explored by Raskar (2013). The experiment derived the plot boundaries from satellite images, and GPS was used for enumeration. The study brought out clearly the advantages of deriving accurate plot-level data.

Attempt has been made to automatically extract the farm field delineation from the satellite images. In a research study by Alemu (2016), useful algorithms were developed and tested for automatic extraction of field bunds using different approaches like single, multiple bands and also texture bands. It was found that results are as expected in areas with heterogeneous landscapes. The study recommended for having good reference datasets and thorough knowledge of the study area.

GPS has been used to trace the crop plot boundaries and enumerate the crops in Masaka District of Uganda (Schøning et al. 2005) as a part of Pilot Census of Agriculture (PCA) experiment. Measurements from GPS and measuring tape were not having any statistically significant errors and hence found to be useful. This experiment is tested and introduced as a cost-efficient tool for area measurement and for geo-referencing of holdings. The study recommended that there is potential to use relatively cheap Global Positioning System (GPS) equipment for measuring of area and for geo-referencing of holdings in the context of agricultural statistics. GPS measurement without additional equipment for adjustments of signals and/or improved antenna proves to give some variation in the measured area. A pixel-based cropland classification study based on the fusion of data from satellite images with different resolutions based on a time series of multispectral images has been conducted by Crnojevic et al. (2014). The experiment was conducted with images from Landsat-8 and RapidEye. The method effectively exploits available data and provides an improvement over the existing pixel-based classification approaches through the combination of different data sources. Another contribution of this paper is the employment of crowdsourcing in the process of reference data collection via dedicated smartphone application.

All the methods using satellite images had the essential component of reference of ground truths. In a mixed cropping pattern with small and fragmented holdings, the extent of ground truths was found to be inadequate. This aspect has led to exploration of use of GPS for complete enumeration where the reliability is unquestioned and every crop would have been mapped and one can get spatial attributes of crops (Naik et al. 2013). The experiment conducted by a consortium of Indian Institutes of Management, Bangalore, has indicated better approach for crop mapping. The crop plot boundaries were traced by handheld GPS and compared with field measurements by tape. The data was registered with village cadastral maps and area statistics was generated. In continuation of the experiment (IIMB and ZOOMIN 2013), the plot boundaries generated from handheld GPS were compared with high-resolution satellite images. Based on the results of GPS tracing of plot boundaries and their registration with satellite image, the study has recommended the use of satellite data to extract the plot boundaries.

The studies conducted so far lead to an inference that manual extraction of field boundaries with thorough knowledge of the landscape provides useful results and provides control on datasets for further validation.

2.3 Crop Mapping Practices in India

The historic references to crop statistics generation in India date back to Kautilya's Arthashastra as well as the Moghul era (Dadhwal et al. 2002; Patil 2012). Currently, the crop statistics are generated based on land revenue system for major food crops and non-food crops. The data is received from the State Agricultural Statistics Authorities (SASA) in various states and union territories (CSO 2008). For every agricultural year (July-June), the Directorate of Economics and Statistics (DES), an attached office of the Department of Agriculture, Cooperation and Farmers Welfare, releases four advance estimates (AE) of the production of major agricultural crops of the country, followed by final estimates. Each of these five estimates is available state-wise and at the national level for the identified crops. While finalizing all India level estimates, the crop-wise data on area, production, and yield received from state governments is thoroughly scrutinized on the basis of information from alternative sources on area, production and yield, rainfall conditions, previous trends in crop-wise area, production and yield in the respective states, commodity-wise trends in prices, procurements, etc.

The system of data collection of crop statistics in the country is dependent on land records maintained for collecting land revenue from farmers. In three permanently settled states (Kerala, Orissa, and W. Bengal), 20% sampling on rotation basis is used, and northeastern states rely on ad hoc surveys. The states and UTs where land records are maintained by the revenue agencies are referred to as "land record states" or temporarily settled states (Andhra Pradesh, Bihar, Chhattisgarh, Gujarat, Haryana, Himachal Pradesh, Jammu and Kashmir, Jharkhand, Karnataka, Madhya Pradesh, Maharashtra, Punjab, Rajasthan, Tamil Nadu, Uttar Pradesh, and Uttarakhand and four UTs of Chandigarh, Delhi, Dadra and Nagar Haveli, and Pondicherry accounting for about 86% of the reporting area). In these states, multi-season full enumeration approach is adopted for generating the crop statistics. Acreage estimates from these surveys have to pass through a hierarchy of aggregation of village, taluka, district, and state level, which contributes to a delay in compilation of national forecasts.

In the states where land records are maintained, the village accountant (*Patwari*) is in charge of a village or a group of villages for carrying out field to field crop inspection in each crop season for an agricultural year to record the crop area and land utilization statistics. He is supposed to record the crop details related to area and land utilization in the Khasra registry (Village Crop Registry—"Girdawari"). After the completion of entries for each survey number of the village, an abstract of area sown under different crops is prepared and sent to next higher official in the revenue hierarchy. Village Crop Abstract (Jinswar statement) is prepared and sent to the next higher official in the revenue hierarchy. At the end of each agricultural year, land utilization area statistics are compiled, and an abstract is sent to related higher official. The crop-wise and land utilization-wise area statistics obtained from different villages are aggregated at the revenue circle, tehsil and district levels. The district-wise area statistics are sent to the State Agricultural Statistics Authority (SASA), which is generally Director of Economics and Statistics or the Director of Agriculture or the Director of Land Records. The state-level aggregation is done by SASA and forwarded to the Directorate of Economics and Statistics (DES), Ministry of Agriculture and Cooperation, Government of India, which is the nodal agency for releasing the state-level and the all India level estimates.

With an objective of improving the quality statistics, the National Sample Survey Organisation (NSSO) introduced a scheme "Improvement of Crop Statistics Scheme (ICS)" through joint efforts of center and state authorities. In this scheme, NSSO carries out the supervision and physical verification of *Girdawari* in a subsample of four clusters of five survey numbers in each of the TRS (timely reporting system) sample villages. An assessment is made for the extent of discrepancies between the entries of a supervisor and *Girdawari* completed by village accountant for each of the selected survey numbers in the sample. The supervisors checking possible errors of aggregations also scrutinize the crop abstract of the village, which is prepared by *Patwaries*. The permanently settled states are also covered under this scheme where a subsample of Establishment of an Agency for Reporting Agricultural Statistics (EARAS) sample villages (survey number) is scrutinized following the same methodology adopted for temporarily settled area. Generally, a total of 10,000 sample villages are covered by the ICS and out of which 8500 are in the temporarily settled states and 1500 in the permanently settled states.

Different organizations/departments are involved in the collection of agricultural statistics, and there seems to be lack of uniformity and coordination among themselves. This needs to be reorganized within the framework of administrative and socioeconomic setup, which will definitely make the whole system more reliable and effective. Agriculture is a dynamic sector where cropping practices change over time and new crops especially of short duration are sown and harvested. However, the Village Crop Registry (Khasra Registry) and other records maintained by *Patwari* have remained almost the same since long. The list of crops covered by the Village Crop Abstract (*Jinswar*) needs a review that may also result in some changes in the manual of instructions for the *Girdawari*.

The Government of India has started on an operational basis a major scheme "FASAL" as an alternative source of data for selected crops (*State of Indian Agriculture* 2015) on regular operational basis remote sensing has also been utilized for crop statistics in India. A central sector scheme—"Forecasting Agricultural output using Space Agro-meteorology and Land based observations (FASAL)"—is implemented with partnership of India Meteorological Department (IMD), New Delhi; Space Applications Centre (SAC), Ahmedabad; and Institute of Economic Growth (IEG), New Delhi. Under the scheme, the release of multiple in-season forecasts is envisaged at the national, state, and district levels based on remote sensing, Agromet, and econometric models in respect to 11 major crops: (1) rice (*Kharif* and *Rabi*), (2) *Jowar* (*Kharif* and *Rabi*), (3) maize, (4) *Bajra*, (5) jute, (6) *Ragi*, (7) cotton, (8) sugarcane, (9) groundnut (*Kharif* and *Rabi*), (10) rapeseed and mustard, and (11) wheat.

Under the revised strategy since 2011, the operationalization of crop forecasts and drought assessment through remote sensing methodology developed by ISRO are being done by the Mahalanobis National Crop Forecast Centre (MNCFC), DAC&FW. Presently, the MNCFC provides forecasts based on remote sensing technology in respect to eight crops, viz., (1) rice (*Kharif/Rabi*), (2) wheat, (3) rapeseed and mustard, (4) cotton, (5) jute, (6) sugarcane, (7) *Jowar*, and (8) potato. The forecasts generated by the MNCFC are based on the yield models developed by the IMD, and the area coverage is based on remote sensing technology.

Chapter 3
Procedure Adopted in Different States

Abstract Generation of crop area statistics in Karnataka State is a collaborative process involving different departments. At state level, the Department of Economics and Statistics processes the reconciled crop area reports submitted by the districts. Primary process starts at a particular village, and the data gets aggregated at different levels of administrative hierarchy. At the village level, statistics on land use, crop area, and crops irrigated are collected through a joint field inspection by a team comprising of Village Accountant, Agricultural Assistant, and Work Inspector from the Water Resources Department. In Andhra Pradesh, Village Revenue Officer at village level is mainly responsible for the collection of land utilization and crop statistics information every season with the help of Village Revenue Assistants which is verified and compiled and monitored at Mandal level by Mandal Revenue Officer. The data are further reviewed at district level and accepted by the Directorate of Economics and Statistics at state level. In Odisha, Directorate of Economics and Statistics (DES) is responsible for collection of information on land utilization and crop statistics every year. The District Planning and Monitoring Units (DPMUs) undertake village-level surveys in approximately about 20% villages covering each Block/Mandal every year. The state has formulated *EARAS* (Establishment of an Agency for Reporting Agricultural Statistics) framework defining coverage of villages for every five years. Over a period of five years, all the villages in each Block/Mandal are covered through EARAS randomized sampling framework.

Keywords Karnataka · Andhra Pradesh · Odisha · Directorate of Economics and Statistics · Land use · Village Accountant · Village Revenue officer, EARAS

Each state in the country has its own process of generating crop statistics though complying with the standards set in by the national government. The processes adopted in each state are described in the following sections.

3.1 Karnataka

Generation of crop area statistics in Karnataka State has been a collaborative process involving different departments. For generating the advance estimates of crop area statistics, government has been refining the process, and as per the latest guidelines (Government Notification No. RD 23 ELR 2004 dt. 06.05.2005), the procedure followed is as follows.

Primary process starts at a particular village, and the data gets aggregated at different levels of administrative hierarchy. At the village level, statistics on land use, crop area, and crops irrigated are collected through a joint field inspection by a tripartite team comprising:

- Village Accountant (VA, *Patwari*) of Revenue Department
- Agricultural Assistant (AA) of Agriculture Department
- Works Inspector (WI) of Water Resources Department (for major medium irrigation surface sources and minor irrigation surface sources and groundwater sources)

As there is no staff of Horticulture Department working at the village level, the VA of Revenue Department is entrusted with the responsibility of collecting the area survey/sub-survey number-wise of horticulture crops. Similarly, wherever Work Inspectors are not posted/not available, the services of the WI of Water Resources Department (for major, medium, and minor irrigation) available in the surrounding place or any other authorized by the Assistant Executive Engineer are utilized. For minor irrigation tanks having command area of less than 40 ha, WI of Zilla Panchayat (if available) assists the WI of Water Resources Department. Else, it is the responsibility of Zilla Panchayat divisions and sub divisions to give the details pertaining to minor irrigation schemes under their control, to the WI of Water Resources Department (for major, medium, and minor irrigation).

The team for each village within the VA circle is constituted by the Tahsildar in consultation with his counter parts of other departments. A complete list of village-wise teams is maintained by the Tahsildar each year.

3.1.1 The Process

Based on the cropping seasons, calendar of events for enumeration of crop area, time schedule for compilation, and reporting of area have been prescribed (Table 3.1). A process flow chart for generating crop area statistics as per the schedule and based on the guidelines is shown in Fig. 3.1.

Table 3.1 Calendar for crop enumeration, Karnataka State

Season	Enumeration period	Submission of VCS-II to Tahsildar	Submission of VPCS, HCS (for irrigated crops) to Tahsildar	Forward of data to DSO	From DSO to DES
Early *Kharif*	July	10-Aug	20-Aug	31-Aug	30-Nov
Late *Kharif*	Sept		20-Sept		
Total *Kharif*		20-Oct	30-Oct	15-Nov	
Rabi	Jan	10-Feb	20-Feb	01-Mar	10-Mar
Summer	April	10-May	15-May	20-May	25-May

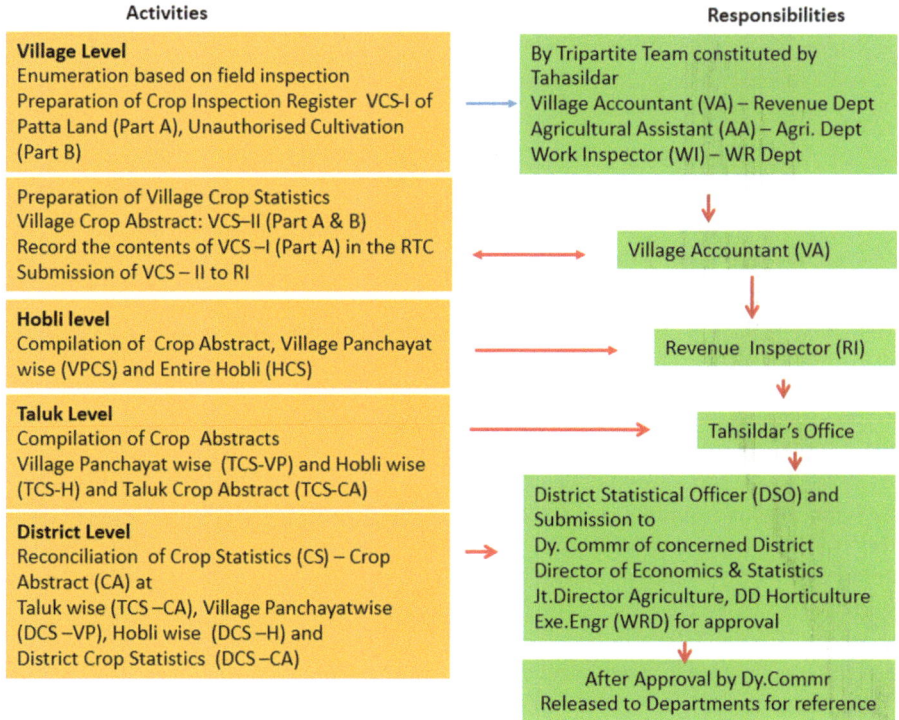

Fig. 3.1 Current process adopted for crop statistics in Karnataka

3.1.1.1 Village Level

The team undertakes field-to-field visit for crop area enumeration as per time schedule prescribed (Table 3.1). Survey/sub-survey number-wise area enumerated on field inspection by this team is compiled by the VA and reported to the Revenue Inspector of the concerned hobli. Following are the records/reports prepared based on the field enumeration:

- Village Crop Statistics (VCS)-I: VCS-I is essentially the primary crop register (*Girdwari*) at the village level. It has two parts as follows:

  - Part A is the Crop Inspection Register of Patta Lands.
  - Part B is the Crop Inspection Register of unauthorized cultivation.

  VCS-I (part A and part B) also contains the breakup of crop areas according to "irrigated" (source-wise under major, medium, and minor irrigation schemes and also the crops irrigated under springs, pickups, seepage water, nalas, ponds, drain water, etc.) and "unirrigated" category.

- Village Crop Statistics (VCS)-II: VCS-II is the abstract of crop pattern (*Jinswar*) which has two parts, viz., part A, crop abstract (CA) of Patta Lands, and part B, crop abstract of unauthorized cultivation.

The VA invariably records the survey number-wise cropped area reported in VCS-I (part A) in the RTC. Functionaries of Agriculture, Horticulture, and Water Resources Department (including major, medium, and minor irrigation) also take care that the actual reconciled crop-wise area figures of their departments are entered in the pahani by the Village Accountants. The area thus recorded in the RTC gets invariably entered in the *Bhoomi*. The VA preserves VCS-I (parts A and B), and copies of reports are sent to the Revenue Inspector in Form VCS-II (parts A and B).

3.1.1.2 Hobli Level

At hobli level, the Revenue Inspector compiles the data submitted by VAs. The Revenue Inspector prepares Village Panchayat-wise report (VPCS) and hobli-wise report (HCS) separately containing village-wise information for each crop. He also prepares VPCS (irrigation) and HCS (irrigation) separately containing village-wise, source-wise irrigated area on the basis of VCS-I (part A). The information in the reports VPCS and HCS is attested by the Revenue Inspector and Agriculture Officer, and the information in the reports VPCS (irrigation) and HCS (irrigation) is attested by the Revenue Inspector and Junior Engineer (minor irrigation), Water Resources Department, and then sent to the Tahsildar.

3.1.1.3 Taluk Level

At taluk level, the hobli level reports, i.e., VPCS and HCS and VPCS (irrigation) and HCS (irrigation) reports received from the RIs, are scrutinized and processed, and the taluk level reports, i.e., TCS-VP containing Village Panchayat-wise information and TCS-H containing hobli-wise information on crop area, are prepared. On the basis of TCS-H, the taluk crop abstract (TCS-CA) is prepared. The information contained in VPCS and HCS and VPCS (irrigation) and HCS (irrigation) is used for the preparation of annual crop statistics report.

Taluk report containing both part A and part B, i.e., TCS-CA, is approved by the committee consisting of the following officers:

1.	Tahsildar	Chairman
2.	Assistant Director of Agriculture	Member
3.	Senior Assistant Director of Horticulture	Member
4.	Nodal Officer of Water Resources Department	Member
5.	Statistical Inspector stationed at Tahsildar's office	Convener

The taluk crop abstract (TCS-CA) is sent to the Deputy Commissioner and the District Statistical Officer (DSO) for each season as per time schedule specified.

The data received from village level and hobli level are computerized at the Taluk office.

3.1.1.4 District Level

At district level, the reports (TCS-VP, TCS-H, and TCS-CA) sent from Tahsildars are reviewed and compiled to prepare district reports, viz., DCS-VP containing Village Panchayat-wise and taluk-wise information, DCS-H containing hobli-wise and taluk-wise information, and DCS-CA containing crop-wise abstract of the district.

District report, i.e., DCS-CA, is approved by the committee consisting of the following officers:

1.	Deputy Commissioner	Chairman
2.	Joint Director of Agriculture	Member
3.	Deputy Director of Horticulture	Member
4.	Nodal Officer of Water Resources Departments	Member
5.	District Statistical Officer	Member Secretary

After the approval of the reconciled crop-wise area by the Deputy Commissioner, the crop area figures are released by the concerned departments.

3.1.1.5   State Level

At state level, the Department of Economics and Statistics (DES) processes the reconciled crop area reports submitted by the DSO, and the state level report is prepared. This report is the authoritative source on statistics of crop area.

3.1.2   Issues

The traditional system for crop area enumeration makes use of Village Accountants (along with other supporting/supervisory personnel) for field enumeration for a particular geographical (administrative unit) area. The main constraints within the system are as follows:

- Shortage of staff and more villages for a VA: Due to shortage of VAs, on an average about five villages are handled by a single person. Further, VA is loaded with other competing responsibilities, and his focus is lost.
- VA does not have the latest/updated village maps and cannot cover the allotted geographical area in view of other activities imposed on him.
- The RTC records are not updated, and VA does not have a print of RTC register which is used for updating the crop information.
- The crop area details are recorded based on the information provided by the concerned farmer(?)/duplicate of earlier records.
- Due to certain requirements, farmers may have reported the crop as well as area of his/her holding and that gets recorded in the RTC as there is substantial variation in the crop physically observed and that available in the records.
- Individual survey numbers are normally divided, and sub-survey numbers are depicted as individual holdings, and for each holding, *Khata* register exists. But, this is not updated in the RTC.
- Accurate survey boundaries and drawings are not integrated. Only a composite village maps showing survey numbers is available.
- Village Accountants have not studied completely the details of his/her jurisdiction particularly the information related to number of holdings, pattern of holdings, and information pertaining to government lands.
- Eyeball estimates that are required to be done may not be accurate.

3.2   Andhra Pradesh

Village Revenue Officer (VRO) at village level is mainly responsible for collection of land utilization and crop statistics information every season. VRO undertakes this exercise with the help of Village Revenue Assistants (VRAs). VRO reports this crop statistics/land utilization information to Revenue Inspector (RI) at Mandal level during beginning of every crop season. This will be cross verified by Mandal-level Assistant Statistical Officer (ASO) along with Agricultural Officer (AO) and Assistant Engineer (from Irrigation Department). After cross-verification and joint approval at Mandal level (RI, ASO, AO and AE), this information is submitted to Mandal Revenue Officer (MRO). He recommends this information to Revenue Divisional Officer (RDO) at divisional level. Upon approval from RDO, it moves to District Revenue Officer (DRO). He further forwards to Collector at district level. Similarly, this information is also shared from ASO at Mandal level to Deputy Statistical Officer at division level. He/she forwards to Statistical Officer (SO) at district level (Fig. 3.2). The SO finally with his recommendation forwards to CPO (Chief Planning Officer) at district level. CPO further reports this information to Directorate of Economics and Statistics at state level. Traditionally, VRAs used to collect the field level information using standardized forms. Recently, the Government of Andhra Pradesh has introduced "e-crop" application where the VRAs can feed the information through online login credentials. Overall, the process of data collection and involvement of key officials at various stages are depicted in the diagram below:

3.2.1 Issues

– Understaff at many Mandals (having responsibility for 3–4 villages).
– Heavy work load (government surveys, protocol visits, official meetings, tele-conferences, etc.).
– Lack of sufficient financial resources.

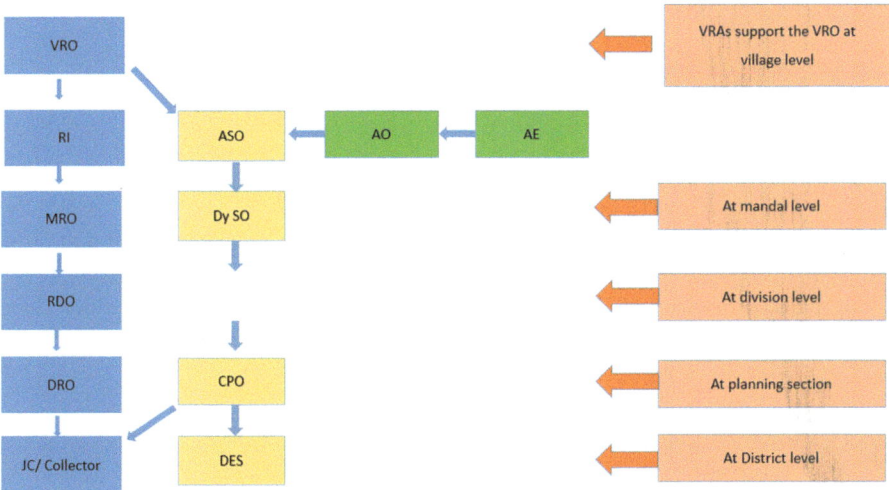

Fig. 3.2 Process for crop statistics in Andhra Pradesh

Table 3.2 Calendar of events for crop enumeration, Andhra Pradesh

Calendar	Start	Reporting by
Kharif season	July/Aug	Sept/Oct
Rabi season	January	Jan/Feb
Summer season	15 March (no crop)	30 June (no crop)

- Involvement of surveyor is necessary; otherwise obtaining accurate information on land utilization status and crop enumeration may not be possible.
- Lack of coordination between Agriculture and Irrigation Departments while estimating crop statistics and sources of irrigated area.
- Manual data collection (plot by plot) is a more cumbersome process. E-crop application of the Revenue Department is good, but it does not capture the GPS coordinates of the location.
- Few VRAs are highly knowledgeable and can adopt new technologies like usage of TABS and e-crop applications.
- ASO is not able to get timely reports/information from line departments.
- Quality of information submitted by line departments is poor.
- Lack of active involvement of other line departments.
- Too much work load and lack of enough time/support staff to validate the information.
- Inadequate financial resources to engage additional field investigators.
- Usage of ICT applications is still limited.

In general, the quality and timely submission of crop statistics/land utilization information lies (80%) with VRO at village level. Concern RI (20%) and MRO (10%) in the Mandal have limited responsibility in terms of cross-verification of the provided information with respective line departments. In many cases, only the extreme outliers will only get rectified or corrected; otherwise it will be a routine process (Table 3.2).

3.3 Odisha

The Directorate of Economics and Statistics (DES), Bhubaneswar, is an apex body of state, responsible for the collection of information on land utilization and crop statistics every year. The Directorate in-turn will coordinate with the District-Level Planning and Monitoring Units (DPMUs) located at each district in the state. They undertake village-level surveys in about 20% villages covering each Block/Mandal every year by involving their staff. The state has formulated *EARAS* (*Establishment of an Agency for Reporting Agricultural Statistics*) framework which defined coverage of villages for the next 5 years (2017–2022) and well-stipulated guidelines to be adopted over a period of time. The staff uses the standard questionnaire forms for collecting the land utilization data from the field. The survey team uses survey number-wise maps for the collection of land utilization information. In general, these surveys are led by an experienced surveyor and with involvement of few field staff. Thus, over a period of 5 years, all the villages in each Block/Mandal are covered through EARAS randomized sampling framework. Block/Mandal is the primary sampling unit in the state rather than the village. So, most of the information is maintained at the Block/Mandal level rather than at village level. The surveyed information collected at each Block level is directly uploaded to DES website for their aggregation.

Senior Investigator is the prime person responsible at district/DPMU level for estimation of crop area coverage and land utilization survey. He/she is supported with eight Block-level Statistical Field Inspectors (SFIs) for implementation of various activities. Each Block-level Statistical Field Inspector in-turn is supported with Statistical Field Surveyors (SFSs) at the village level. They actually monitor all crop surveys in selected villages as per EARAS selection. VAW (Village Agricultural Worker) is the village-level person who coordinates all the department of agricultural activities/surveys (Fig. 3.3).

The EARAS manual (2016–2017 to 2020–2021) was well prepared with selection of Gram Panchayats listed by year-wise. Approximately 20% of villages spread over all Gram Panchayats in a selected Block are randomly selected for undertaking the field survey. The village-wise MAPS (based on settlement of 1992) and ROR (record of rights) are the key instruments used in the field surveys. Standard questionnaires (Form 1, 2, and 2A) are used for collection of crop-based (primary/non-primary) information. During the process, randomly few survey nos. are also selected for organizing "crop cutting" experiments. These crop cutting experiments

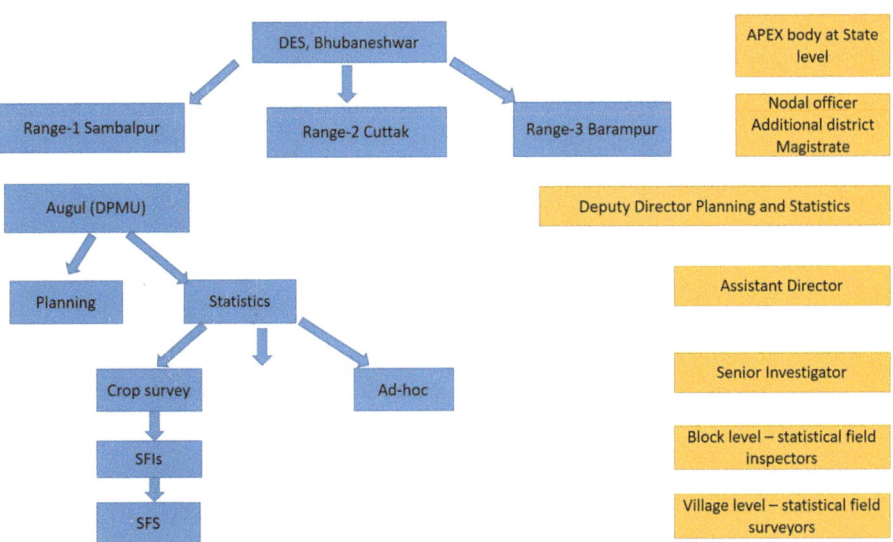

Fig. 3.3 Process for crop statistics in Odisha

are jointly organized and monitored by both Agricultural and Revenue Departments. The "CC App" was in usage for monitoring the crop cutting experiments which is GPS based.

Similarly, the Department of Agriculture also has similar hierarchical structure for undertaking different agricultural development activities in the state. VAW (Village Agricultural Worker) is the prime contact at village level. Agricultural Officer (AO) is the prime person at Gram Panchayat level. Assistant Agricultural Officer (AAO) is the highest officer available at Block level. District Agricultural Officer (DAO) is the main focal person responsible at district level.

Traditionally, the Department of Agriculture never undertakes any crop estimation surveys in the state. The department uses the "eye estimate" method for generating the cropped area under different crops. Specifically, during 2016–2017, the Baranpal Mandal was identified with 32 Gram Panchayat villages to be surveyed as per the EARAS guidelines prepared by DES. But, due to lack of sufficient staff with DPMUs, both agricultural and revenue department staff were engaged in this activity for the first time. Only very limited training was provided to both Agricultural and Revenue Department staff – for undertaking these surveys. The selected 32 Gram Panchayats were distributed among the three departments – Agriculture Department (13), Statistical Department (12), and Revenue Department (7). Additional district magistrate located at district headquarters will act as a "Nodal Officer" for coordination and implementation of various activities initiated by DPMU. There was no involvement of groundwater department engineers/staff in the estimation of irrigated/dryland areas in the villages. The DPMU maintains these records (on hard copies) for at least 10 years for further verification in future. The data is entered in well-designed data entry formats prepared in CS-Pro software.

3.3.1   Issues

– Since 1986 the DES in the state is alone responsible for implementing this procedure for collection of crop statistics/land utilization information. Recently (2016–2017), the Department of Agriculture and Revenue was engaged in the process due of lack of sufficient staff. Despite all efforts, the manpower is inadequate.
– The maps referred by the department are very old; recent structures/ownership details are not incorporated/updated. The land records/tenant lists are very old.
– Application of ICT tools/technologies is very limited.
– Every year only 20% Gram Panchayat villages in the Block/Mandal are covered through field surveys, and the rest of 80% was extrapolated.
– Limited interest and cooperation from villagers.
– Heavy work load because of responsibility of crop insurance lies with DES.
– Always try to minimize the error in paddy crop area estimation at Block level. High importance is given to paddy than other crops. The margin of error in reporting other crops will be relatively higher.

3.4   Overall Issues with Crop Area Statistics in the Three States

Review of the existing procedure for collecting the crop area information indicated that the data being generated at microlevel does not correlate to the ground realities. Geographical distribution of crops is not forthcoming, and errors creep in at all stages, and reliability of the statistics is doubtful (Table 3.3).

Table 3.3 Overall issues with crop area statistics in the three states

Particulars	Guidelines (government specifications)	Actual practice	Gaps in the process	Reasons for gaps
Crop area	Should be based on Village Crop Register – compiled by enumeration of each holding in the field by the concerned official (VA)	• Older Records copied with minor/no manipulation • VA/Patwari hardly visits the village/field	Actual field visits are not done Little OR no consultation with farmers No technology intervention	VA is provided with more villages due to shortage of manpower VA is involved in many activities that hardly enables him to spare time for recording the crop area
Plot boundaries/ holdings	Should have used latest/updated records	No one uses the actual map for enumeration	Village maps not updated with respect to changes VA is not provided with RTC/Khasra Register Many VAs are not aware of the actual boundary of a village Knowledge of VA with respect to Geographical Appraisal and importance of RTC for crop area is limited	VA is not provided with latest maps of villages VA is not provided with RTC register for recording the crops information RTC also does not contain the crop information VA hardly finds time to visit every village in his jurisdiction
Land use pattern	Should be compiled based on field enumeration	Old information is repeated with/ without minor changes	Changes in the land use are not updated either in the table or map	There is no practice of preparing land use map by the concerned office Data is compiled from old records
Boundaries of forest, wastelands, tank area, tank bed	Actual revenue records should have been referred to and data updated	Old information is simply copied without any relevance to site conditions	Village maps are not updated with respect to these features	Concerned authorities may not get time to update these information

Chapter 4
Theme of the Crop Inventory Experiment

Abstract Agricultural statistics should correspond to ground situations and determine the actual extent of each crop along with cultivation and other farming practices. The process should provide reliable, retraceable geographical data that are climate independent and can be applied to different agroclimatic regions. Geo-stamping approach, an integration of remote sensing, GPS, and GIS, has an inbuilt tracking of workflow of any regular data collection and reporting system and allows plot-level inventory. The process is cost-effective, and the basic reference data system can be used repeatedly without much effort. GPS-based smartphones are the best instruments that can be operated by anyone because it provides authenticity of data and captures actual ground situation.

Keywords Farming practice · Geo-stamping · Remote sensing · GPS · GIS · Plot level inventory

Different approaches are adopted in various parts of the world for generating the crop area statistics. However, it is evident from the literature that crop area information is not available before the harvest. The method adopted by the concerned departments in the country has constraints as follows:

- Enumerators do not visit the field during the specified time for inventory.
- Enumerators tend to use the old information (previously available) and pass for further compilation.
- Data compilation happens at different levels, and it takes time for aggregation and errors, if any does not get noticed.
- The departments do not deploy adequate human resources for the work (understaffed as revealed during the discussion with departments).

Therefore, theme of the current research has been exploring means of generating reliable information on crops during the growing season. The literature survey and the analysis of available data indicate gaps in the system, and technological interventions so far have also been not able to produce complete information on crops and their extent. In this context, the following hypothesis has been postulated, and experiments were conducted:

- The statistical data on crops derived by means of extracting the old information from the revenue records does not match with ground situations.
- Current pattern of crop statistics does not provide information on all types of crops grown in any season.
- Geographical context of crop statistics provides microlevel information related to crops, and holding level information is essential to understand the constraints of cultivation and other farming practices.

4.1   Geospatial Technology Intervention for Crop Area Statistics

Geospatial technologies have been recognized as the tools that strengthen the quality of geographical data both in terms of reliability and in built capacity of traceability. The use of satellite images have also been explored for crop mapping, and it is found that regions with fragmented holdings and mixed cropping system, remote sensing techniques have certain limitations. Cloud coverage is one of the factors that hinder the use of remote sensing, and further, in a mixed cropping system, it becomes difficult to delineate different crops. The use of handheld GPS also might pose certain constraints due to signals from orbiting GPS satellites.

Therefore, having reviewed various methodologies for crop area mapping ICRISAT has envisaged to adopt an integrated method for crop enumeration in different agroclimatic regions of the country. The following sections briefly describe the research methodology.

4.2   Geo-stamping Approach

Geo-stamping refers to visualizing WHERE, WHAT, and WHEN an activity happened. It means all the data collected automatically includes location information that can be used for visual map-based reports. The combination of spatial data and locations of activities can be reported to help improve accountability and management insight. When applied to crop cultivation, it allows knowing where and what types of crops are being cultivated and visualization of changing trends.

It has an inbuilt tracking of workflow of any regular data collection and reporting system. Geo-stamping approach (location stamping of each transaction) makes capturing this location information automatic and precise; the crop area data collection system will gain critical insight without additional effort by field personnel. As well, the entire program will be able to achieve:

- Better monitoring of efficiency of process
- Better and effective recording of crops
- Better reporting of location-specific details and eventually predict outcomes

4.2.1 Inventory at Plot Level

Any land holding would have a number of smaller plots, and generally farmers cultivate crops of their choice in each plot depending on local agroclimatic conditions and their economic status and sources of water. Traditionally, the plots do not change unless farmers shift toward changing the entire land use – plantation, non-agricultural use, etc. Therefore, plot-wise inventory would lead to reliable information at holding level.

4.2.2 Generation of Base Information

For conducting the plot-level inventory, a baseline information of the crop plots is essential. These can be conveniently generated from the freely available high-resolution images like Google Earth. The crop plots are clearly discernible in the high-resolution satellite images. The village maps containing survey numbers can be digitized and registered with satellite image so that the crop plots are correlated broadly with the revenue administration units/the holding of individual farmer.

4.2.3 Use of Smartphones

The GPS-based smartphones have an option to load maps that can be used not only for navigating but also registering the spot coordinates. The base information in the form of geo-referenced maps can be converted in to geo-fenced map tiles and loaded. An application can be developed to use the maps only in the geographic region of the village for which the geo-referenced maps and also for display of location on the map loaded. Application can be developed to capture the relevant information on the spot and along with the photographs. Entire data is locally stored in the smartphone itself and later can be transmitted directly to data center. Storing the data on phone is to avoid the constraints of Internet availability at the remote locations in villages.

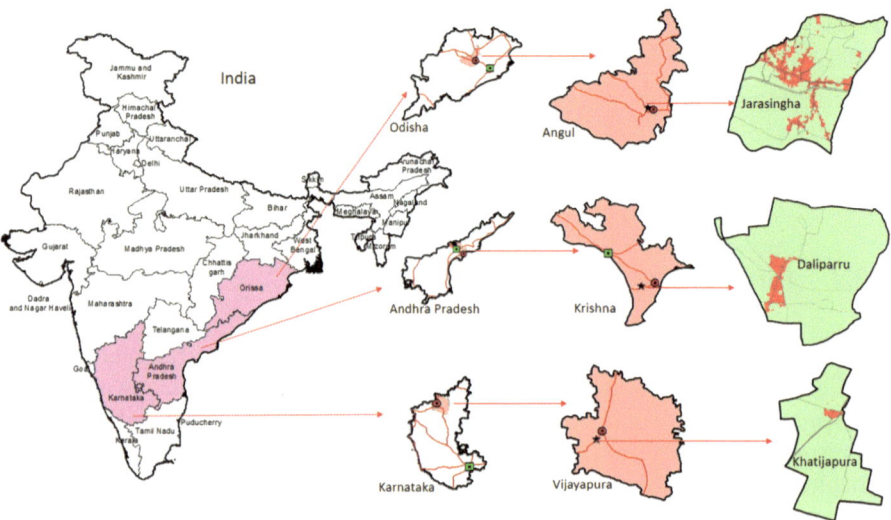

Fig. 4.1 Location of the villages selected for the study

Table 4.1 Villages selected for the research study

Sl.	Village	State	Geographical area
1	Khatijapura	Vijayapura Taluk, Vijayapura District, Karnataka	4.78 km^2
2	Daliparru	Gantasala Mandal, Krishna District, Andhra Pradesh	5.81 km^2
3	Jarasingha	Baranpal Mandal, Angul District, Odisha	3.67 km^2

4.3   Crop Inventory in Different Locations

Having designed the workflow of the project related to crop inventory, ICRISAT conducted the experiments in three different agroclimatic regions of the country India. As a part of the study, one village in three states (Fig. 4.1; Table 4.1) has been selected in consultation with respective state agencies.

Geo-stamping approach has been adopted for generation of crop area statistics of Khatijapura, Daliparru, and Jarasingha villages. The strategic interventions have been usage of high-resolution satellite data for extracting crop plots, GIS for spatial data integration and visualization, GPS for recording relevant details of crop and sources of irrigation, etc. on the spot. The details of the experiment are provided in subsequent sections.

Chapter 5
Development of a Spatial Reference (System) Database

Abstract Field crop plot boundaries can be extracted from high-resolution images with spatial resolution less than 2.5 m from satellites like Cartosat, QuickBird, etc. The village maps along with survey numbers obtained from respective states' departments geo-referenced using control points like natural features, roads, etc. and the coordinates obtained from DGPS were used to increase the positional accuracy. Crop plots were extracted from high-resolution natural color composite images from Google Earth. A controlled mosaic of crop plots for each village developed in GIS environment was converted to geo-fenced map tile and loaded on to smartphone powered with suitable application for automatic navigation and recording the information along with photographs in the field. Farmers in each village were informed about the process and cooperation was sought. Field inventory carried out in Rabi was kept as basic reference of crop plots of different farmers. Android application was modified to reduce the mapping of plots without crops in subsequent season. The data generated in each season was systematically organized in QGIS for visualization and spatial analysis. Farmers not only cooperated but also felt that the process is useful so that they can have reference of crop area and type.

Keywords Plot boundary · High resolution satellite data · Village maps · Controlled mosaic · Geo-fenced map tile · Automatic navigation

5.1 High-Resolution Satellite Data

Taking the reference of earlier work by IIM, Bangalore, it was decided to explore the possibilities of extracting crop plot boundaries from high-resolution satellite data. Therefore, ortho-rectified Cartosat image (Picture 5.1) was tested and used for extracting the plot boundaries. It was found that a sensor with more than 2.5 m pixel resolution is useful as one can distinguish the regular shapes of the field bunds (Picture 5.2). The same exercise was also carried out on Google Earth images. Interpretation and delineation of bunds becomes easy since the natural color composite image (combination of multispectral bands) is provided on Google Earth.

© The Author(s), under exclusive licence to Springer Nature Switzerland AG 2019
K. V. Raju et al., *Geospatial Technologies for Agriculture*, SpringerBriefs
in Environmental Science, https://doi.org/10.1007/978-3-319-96646-5_5

Picture 5.1   Cartosat image of a different area used for testing

Picture 5.2   Field bunds discernible on Cartosat image (yellow lines—field bunds)

Picture 5.3   Google Earth image used for one of the villages of the study

From the exercise it is evident that satellite image without any cloud cover can be conveniently used for extraction of crop plots. If the data acquired during non-crop season is available, the field bunds can be extracted with more clarity. Since the images without cloud cover is a must and that was available (Picture 5.3), current exercise has used Google Earth image.

Historical data from Google Earth was reviewed, and though few sets with cloud-free images were available, the most recent one (acquired during 2016) was chosen (Picture 5.3) to extract the crop plots. Revenue survey maps (cadastral maps) for each of the villages were downloaded from the official websites of concerned departments of respective states. The cadastral maps were converted in to GIS format and superimposed on the high-resolution image from Google Earth. Adequate ground control points were established using DGPS in each village (Picture 5.4) as per the standard operating procedure (SOP). The satellite image along with village map was geo-rectified (Picture 5.5) with the help of DGPS control points for each of the villages (Tables 5.1, 5.2, and 5.3).

5.2   Geo-referencing of Village Map

The village maps obtained from respective states' departments were digitized. The vector data of village map was later geo-referenced using the toposheet (natural features, roads, etc. are used as control points). The geo-referenced village map (along with survey numbers) was overlaid on the Google Earth images (online). Geo-coordinates obtained from DGPS survey were also referred to increase the accuracy.

Base Station   Control Point

Picture 5.4 Establishment of DGPS control points

After ensuring proper overlay of village vector on Google Earth images, the plot
boundaries (field bunds) within a survey number were extracted based on distinct
image characteristics. Visual interpretation was followed to distinguish and identify
the crop plots. Since crop plots have distinct image characteristics and appear as
linear lines and have different tones and texture, they stand out clearly on the image.
The individual polygons based on the image elements like tone, texture, pattern, and
association have been vectorized on the screen itself. The rectified village map
superimposed on the image was used as control for extracting the crop plots. The
digital data (vector derived from Google Earth) then converted in to shape files with
survey numbers and crop plots. Each crop plot was given unique number. The shape
file in UTM projection was later converted to geo-referenced map tiles using tools
in ArcGIS along with all relevant information like plot ID. The revenue records of
each farmer in respective village was obtained and correlated to generate broad
Khata/ holding for further use in the field.

A controlled mosaic of crop plots for the entire village (Picture 5.6) was developed in GIS environment, and geo-fenced map tile for each village was generated.

5.3   Software Applications for GPS

A comprehensive Android application was developed, and the data collection format decided by ICRISAT research team was implemented in addition to standard
field mapping parameters such as crop and land use. The geo-fenced map tile was
loaded to GPS-based smartphones (Picture 5.7), and the application was tested for
ease of data authentication and recording of geo-coordinates of each crop plot and

Picture 5.5 Geo-rectified satellite image + village map used for extraction of crop plots

recording the information on the spot along with photographs. The application has the following functionalities:

- Automatic loading of the village map tile on the screen in the field and displaying of location of the enumerator.
- Capturing of the location (coordinates) of the plot (registering the plot).
- Recording details of the crop in the plot.
- Capturing of photograph of the crop.
- Capturing the coordinates of features like borewells and any other structures in the land.

Table 5.1 DGPS coordinates, Khatijapur Village, Vijayapura (Bijapur) Taluk and District, Karnataka

Sl. No.	Pt-ID	WGS84-spherical coordinates		UTM-projection coordinates	
		Latitude	Longitude	Easting	Northing
1	GPS1	20° 51′ 19.34111″ N	85° 04′ 10.33120″ E	299148.020	2307346.386
2	GPS1A	20° 51′ 19.84399″ N	85° 04′ 07.51502″ E	299066.792	2307362.829
3	GPS2	20° 50′ 47.13228″ N	85° 03′ 59.21064″ E	298814.630	2306359.662
4	GPS3	20° 51′ 47.47708″ N	85° 04′ 38.66666″ E	299977.515	2308201.896
5	GPS4	20° 51′ 52.33888″ N	85° 03′ 49.87646″ E	298568.906	2308368.343
6	GPS5	20° 51′ 17.35795″ N	85° 03′ 36.14438″ E	298158.970	2307297.282
7	GPS6	20° 51′ 17.70418″ N	85° 03′ 58.86183″ E	298815.844	2307300.024
8	GPS7	20° 50′ 34.43118″ N	85° 02′ 03.37084″ E	295460.800	2306009.604

Table 5.2 DGPS coordinates, Daliparru Village, Gantasala Mandal, Krishna District, Andhra Pradesh

Sl. No.	Pt-ID	WGS84-spherical coordinates		UTM-projection coordinates	
		Latitude	Longitude	Easting	Northing
1	GPS1	16° 09′ 04.29589″ N	80° 58′ 34.04957″ E	497447.493	1785659.802
2	GPS1A	16° 09′ 01.93061″ N	80° 58′ 34.38919″ E	497457.571	1785587.124
3	GPS2	16° 08′ 54.92948″ N	80° 58′ 24.80419″ E	497172.892	1785372.037
4	GPS3	16° 09′ 36.65202″ N	80° 58′ 01.99532″ E	496495.723	1786654.130
5	GPS4	16° 09′ 52.68234″ N	80° 59′ 08.75092″ E	498478.137	1787146.462
6	GPS5	16° 09′ 31.99516″ N	81° 00′ 03.16028″ E	500093.848	1786510.761
7	GPS7	16° 08′ 29.27039″ N	80° 58′ 52.49365″ E	497995.137	1784583.529
8	GPS6	16° 08′ 37.67225″ N	80° 59′ 34.21992″ E	499234.370	1784841.612

Table 5.3 DGPS coordinates, Jarasingh Village, Banarapal Block, Angul District, Odisha

Sl. No.	Pt-ID	WGS84-spherical coordinates		UTM-projection coordinates	
		Latitude	Longitude	Easting	Northing
1	GPS1	20° 51′ 19.34111″ N	85° 04′ 10.33120″ E	299148.020	2307346.386
2	GPS1A	20° 51′ 19.84399″ N	85° 04′ 07.51502″ E	299066.792	2307362.829
3	GPS2	20° 50′ 47.13228″ N	85° 03′ 59.21064″ E	298814.630	2306359.662
4	GPS3	20° 51′ 47.47708″ N	85° 04′ 38.66666″ E	299977.515	2308201.896
5	GPS4	20° 51′ 52.33888″ N	85° 03′ 49.87646″ E	298568.906	2308368.343
6	GPS5	20° 51′ 17.35795″ N	85° 03′ 36.14438″ E	298158.970	2307297.282
7	GPS6	20° 51′ 17.70418″ N	85° 03′ 58.86183″ E	298815.844	2307300.024
8	GPS7	20° 50′ 34.43118″ N	85° 02′ 03.37084″ E	295460.800	2306009.604

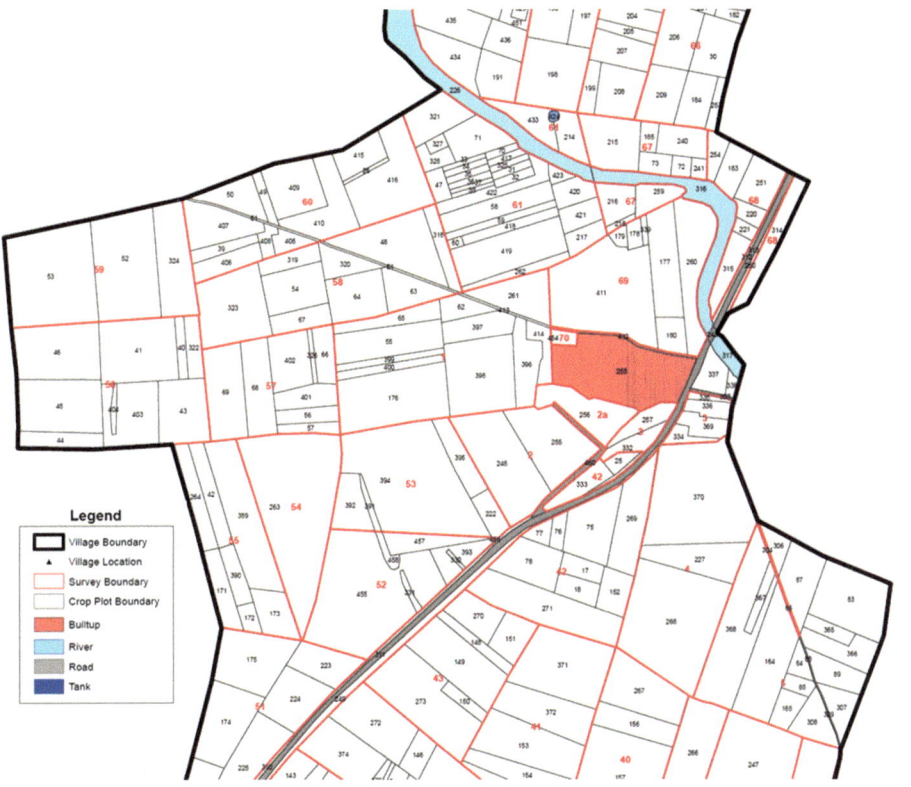

Picture 5.6 Controlled mosaic of crop plots extracted and registered with village map

- Capturing other information (as specified by ICRISAT research team).
- Downloading the data and synchronizing with the GIS database residing in the backend system.
- Separate module/application for enumeration during subsequent season (Picture 5.7).

Description for Survey Number and Holding

1. In a village, for the administrative convenience as well as legal purposes, the lands have been demarcated as survey numbers, and each unit is given a numeric number. Each survey number is further subdivided in to *hissas*/divisions. Individual person may own one or more *hissa*/division. A piece of land (either a *hissa* or division) owned by a particular person is called holding. Therefore, a survey number, depending on the local situation, may have many holdings. Further, a single holding may contain single or more plots. These plots are essentially developed as land/crop management purposes. Accordingly, depending upon the local topographical situations, a holding may contain more than one crop plot.

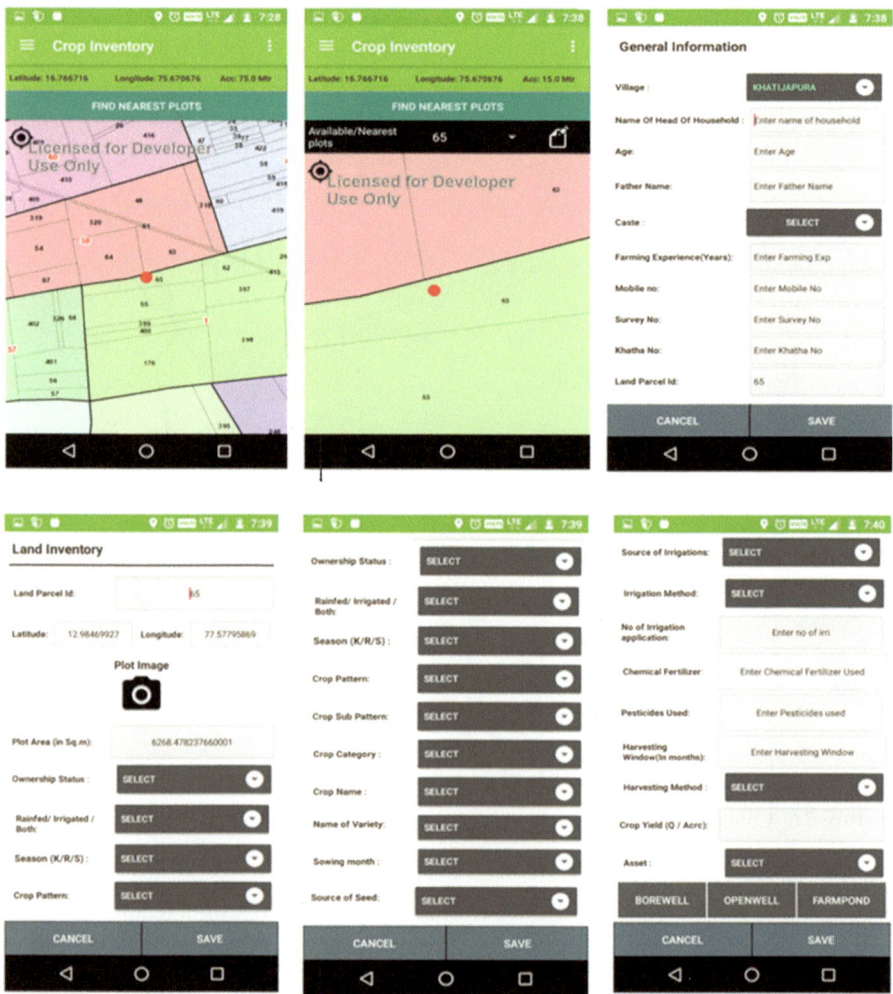

Picture 5.7 Sample screenshots of Android app

2. Holding is actually the land owned by a farmer/person. Each holding has a unique format of information called RTC (record of *r*ights, *t*enancy, and *c*rop details) which describes the extent and crops grown besides other information. The dimensions of holdings vary in each village. Anything less than 1 hectare is called marginal holding and 1 to 2 ha is called small. Similarly, it is common to express the extent of land as size of the holding. Further, a single holding may have plots belonging to different persons, and in such cases it is called fragmented holding (Picture 5.6).

5.3.1   Modification of Android Application

During the first inventory (*Rabi* season), all the crop plots were geo-stamped, and crops were enumerated. Subsequently, the team had detailed discussion with farmers about the cultivation in different seasons. It was learnt that very few farmers grow crops in the summer season. Therefore, geotagging each crop plot becomes redundant. In view of this though field survey was conducted, a process for recording the details has been modified, and the Android application was accordingly modified. Similarly, before taking up inventory during *Kharif* (August to September 2017), discussion was held with farmers, and the practice cultivation was understood. It was clear from the discussion that maximum farmers go for paddy cultivation in Jarasingha and Daliparru and Tur in Khatijapura. In order to have a composite view with reference of earlier season (*Rabi*) crop in each plot and display of plots within a range, the Android application was slightly modified.

5.3.2   Modification for Summer Inventory

The geodatabase of the *Rabi* season was carefully reviewed, and a standard grid of 100×100 m was identified on the village map having all the earlier details. The map was again converted in to geo-fenced tiles and loaded on the smartphone, and the earlier application was modified with a search plots algorithm. Whenever the surveyor visits the field, the application would guide him to navigate to the center of the grid and displays number of plots within 100 m square. Surveyor can observe and identify the plot having crop. If all the plots do not have crops, he needs to record the coordinate of the spot and record the details (as no crop in the plots displayed) and move to the next grid. If any plot has crop, the application alerts him to visit the particular plot and record the details of the crop. The process is repeated for all the grids. This saves time and yet provides reliability of having recorded the field details.

Based on the size of the plots, dimensions of the grid were decided for each of the villages. It was 100×100 m for Khatijapura in Karnataka and 50 m for both Jarasingha in Odisha and Daliparru in Andhra Pradesh. The screenshots of mobile application (Picture 5.8) are shown below.

5.3.3   Modification of Android App for **Kharif** Inventory

In order to have a composite view and display of plots within a range, the Android application was slightly modified. With this, the enumerator would be able to define a range, i.e., 50 m or 100 m from his location. Within the defined range, all the plots (along with unique plot ID) would be displayed (Picture 5.9). Also the crop details

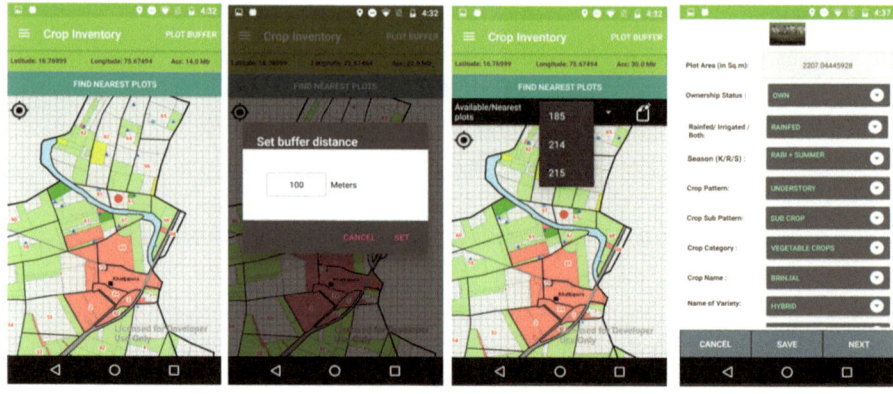

Picture 5.8 Screenshots of the modified application for summer inventory

Picture 5.9 Screenshots of the modified application for *Kharif* inventory

of each plot in the previous season are displayed on the screen of the GPS device. This helps the enumerator to have a clarity on the plots having crops and not having crops. He can avoid moving to the vacant plot and can record details of plots having crops. If all the plots do not have crops, he needs to record the coordinate of the spot and record the details (as no crop in the plots displayed) and move further to a different location to fix a range. In the new spot for a particular range, the plots that were displayed earlier will not be appearing (as details for these were recorded standing at the previous location). This process introduces efficiency of plot enumeration.

Applications for crop mapping were developed using the following tools:

- Operating system—Android.
- Programming language—JAVA in association with Android SDK, Studio.
- Database—SQLite (local database in the smartphone).

5.4   GIS Core and Spatial Database

The vector data of the village survey number and the crop plot boundaries with unique identification number was organized in GIS as per the spatial database convention. The map was configured using WGS 84 map projection system so that the geographical coordinates obtained from GPS for each polygon gets seamlessly integrated.

The GIS database was converted in to map tiles and was loaded on to the smartphones. The field results/crop details and other attribute data captured in the field using Android application with GPS facility were synchronized with GIS database.

The GIS database has further been resolved in to different layers such as mentioned in the following Table 5.4.

Other Preparations

- Holding informations (RTC) for these three villages have been downloaded from the website of respective states as reference for field enumeration.
- Field project team has been given proper orientation and training for crop plot registration with the help of smartphone loaded with special application.
- Data correlation/nomenclature with regard to crops has been tested and validated (back office), and back-office team has been kept in readiness.
- Project implementation team has been apprised about the responsibilities and task of each member.

Table 5.4 Structure of GIS database

Sl. No.	Shape file layer name	Layer type	Attribute of layer
1	Village boundary	Polygon	Polygon vector layer of village boundary containing village, hobli, tehsil, district, state, and area details
2	Assets (borewell, silt trap, farm pond, etc.)	Point	Point vector layer of borewell and silt trap locations and borewell type, year of drilling, depth, yield, and status details
3	Survey number boundary	Polygon	Polygon vector layer of survey number boundary contains survey number, area, village, etc. details
4	Built-up (settlement with households - only vector)	Polygon	Polygon vector layer of individual buildings
5	*Rabi* crop plots	Polygon	Polygon vector data of *Rabi* season crop plots contains 41 attributes
6	Summer crop plots	Polygon	Polygon vector data of summer season crop plots contains 41 attributes
7	*Kharif* crop plots	Polygon	Polygon vector data of *Kharif* season crop plots contains 41 attributes

The spatial database of villages has been compiled and organized in QGIS

Picture 5.10 Discussion with local farmers and officials

5.5   Field Inventory

Before starting the actual field enumeration, a formal discussion with local farmers was held, and the farmers were appraised about the work. In each village, scientists from ICRISAT participated in the meeting and have provided useful coordination with Government Department (Picture 5.10). Respective Village Accountants were called and informed about the initiative and also demonstrated the survey using the GPS. Concerned district and taluk officials were appraised about the project, and cooperation from different officers was sought.

5.5.1 The Field Procedure

Following are the specific activities that were carried out in the field:

- The farmers of particular village were requested to participate in the enumeration process (Picture 5.11). They were specifically told to confirm their plots.
- After confirmation of every plot, the details like name of the crop grown was recorded in the GPS, and photograph of the crop is taken.
- In case of any source of irrigation is seen in the land, the same is captured using GPS.
- Further, details of crop pattern in the field are recorded, i.e., whether it is single crop or crop interspersed with other crop (mixed crop).
- Physical structures like building, sheds, etc. (in case seen in the plot enumerated) are traced.
- Since the Internet connectivity was week, the data recorded in the GPS was stored locally in the GPS itself and was downloaded to a computer at the end of the day's session.

5.5.2 Back-Office Work

- The data received from each GPS unit from the field on daily basis was maintained in separate folders.
- The attribute information of each crop plot was integrated, and a special team for quality checking was deployed for analysis of recording errors, if any, like name of crop.

Picture 5.11 Farmers' participation in enumeration process

Table 5.5 Crop inventory duration

Sl. No.	Village	Field sessions (crop plot enumeration)		
		Rabi season (February 2017)	Summer season (Apr–May 2017)	*Kharif* season (Aug–Sept 2017)
1	Khatijapura	07.02.2017 to 12.02.2017	24.04.2017 to 27.04.2017	16.08.2017 to 18.08.2017
2	Daliparru	09.02.2017 to 19.02.2017	24.05.2017 to 26.05.2017	08.09.2017 to 11.09.2017
3	Jarasingha	07.02.2017 to 26.02.2017	23.05.2017 to 27.05.2017	09.09.2017 to 16.09.2017

- All the data from GPS devises were systematically ported to the basic GIS database organized on the open-source GIS platform, i.e., QGIS. Different layers for specific crops and other base details were created, and the area statistics were generated.

The formats used for data collection are provided as annexure.

Duration of crop inventory carried out in three villages (Table 5.5) is as follows.

5.6   Response from the Farmers

One of the important aspects of this experiment was the interaction with farmer who identifies his/her land and provides information. The responses vary from state to state. Though the crop in any plot is directly identifiable by the enumerator and the plot gets registered in the GPS (with time, date, and geographic coordinate and crop photo), additional information from the farmer is useful to get insights in to the farming practices and use of fertilizers/pesticides, etc. Every farm plot was inventoried during the field session, and information was collected from the farmers using the Android application. The response from farmers in providing inputs during the inventory is summarized in the following Tables 5.6, 5.7, and 5.8.

In Jarasingha village, very few farmers participated in the field session when crop inventory was being carried out. However, for Kitchen Garden interview, farmers responded, and the details are provided below.

Table 5.6 Summary of farmers' response in Khatijapua

Total number of attributes used per farmer in the Android App	% of farmers providing all the details	Reasons for not getting certain attributes
41	97.56	Not able to get mobile numbers of some farmers • Few of them reside away from the village and • Few of them informed not having the number

Total no. of attributes captured in Khatijapura Village, 5592

Table 5.7 Summary of farmers' response in Daliparru

Total number of farmers in the village of Daliparru: 597			
Number of farmers responded/provided details for attributes			
Number of farmers	Number of attributes provided by farmers	% of farmers responded	Reasons for not filling certain attributes
33	41	5.5	
558	40	93.5	Not able to get mobile numbers of some farmers, since they reside away from the village/town, NRIs
6	29	1.0	Not able to get farmer's personal details (name, age, father's name, caste, etc.) and livestock details, since they are residing away from the village and also details missing in RTC (Adangal)

Total no. of attributes captured in Daliparru Village, 23,841

Table 5.8 Summary of farmers' response in Jarasingha

Number of households responded, 206			
Number of attributes used for each household, 57			
No. of household (Kitchen Garden)	Number of attributes responded	% of households responded	Reasons for not providing details
135	57	65.53	Farmers not interested to involve in this activity during survey
71	55	34.47	

Total no. of attributes captured in Jarasingha Village (farmers having Kitchen Garden), 11,600

Chapter 6
Results of Field Inventory

Abstract The three villages located in different agro-climatic regions are conspicuous with respect to sources of irrigation, cultivation practices, cropping pattern in different growing seasons. Khatijapura village in Karnataka is entirely a rainfed village with limited groundwater development. Crops are grown in both kharif and rabi seasons and to a limited extent in summer. Cropping in different seasons is practiced and characterized by small to medium holdings with few exceptions and the plots are fairly large. About 17% of the net sown area is under lease, contract farming and crop sharing practices. Daliparru village located in coastal plains of Andhra Pradesh has source of irrigation for kharif season. Characterized by small to medium holdings, village practices almost mono-cropping—paddy (irrigated) during kharif and pulses during rabi. Having more than 15% non-resident farmers, leased cultivation is predominant. During summer hardly any crops are grown. Jarasingha village, a part of interior coastal midland of Odisha is supported by an irrigation project. Characterized very small to small holdings practices cultivation in both kharif and rabi season while summer season cropping is limited to small extent and groundwater is being used as supplementing source. Crop sharing is practiced during kharif season and leasing is seen during rabi. Kharif is dominated by paddy while in rabi, it is mixed cropping. Kitchen garden practice is conspicuous of Jarasingha.

Keywords Agro-climatic zones · Rainfed · Kharif · Rabi · Summer · Cultivation practice · Lease · Contract farming · Share cropping · Mono-cropping · Small holdings

Plot-level crop inventory was carried out using GPS-based mobile devices in all the three villages. There is substantial variation in the dimension of the crop plots and also cultivation practices. The size of the crop plots indirectly provides insights in to the water management/irrigation practices and also the landholding pattern. During the field survey, information related to few socioeconomic aspects have also been collected from the farmers. Details of individual villages are provided in subsequent sections.

6.1   Khatijapura

Khatijapura village is a moderately large village covering 478.44 ha of land. It is a part of north interior maidan of Karnataka. It is classified as part of Northern Dry Zone (agroclimatic zone). The agriculture is totally dependent on monsoonal rainfall. The village is characterized by small to medium holdings with few exceptions, and the plots are fairly large. About 324 crop plots are inventoried with minimum area of 3 Guntas and maximum of 12 acres 30 Guntas. Average area of crop plot in this village is 3 acres 6 Guntas.

6.1.1   Sources of Irrigation

The source of water for crops is rainfall, and only few farmers have developed groundwater sources for cultivation. Thirteen irrigation borewells are noticed mostly in the northern part of the settlement, where one is in dry condition. The quality appears to be brackish. Few old open wells have been observed to be dry. The yield in the borewells is moderate.

6.1.2   Cultivation Practices

Village is located 10 kms from Vijayapura City, almost 30% of the land owners reside in Vijayapura city, and practices such as contract farming, crop sharing, and leasing out of farm lands are observed. 17% of the net sown area in Khatijapura is under such practices where contract farming is 9%, crop sharing is 6%, and land Leased out is 1% (Fig. 6.1).

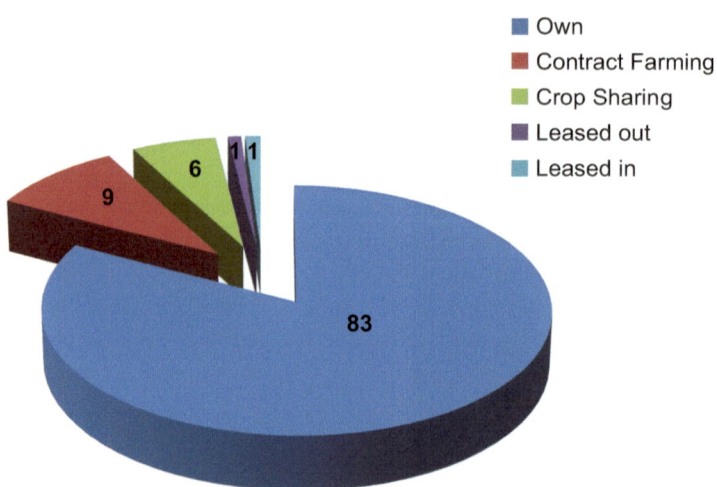

Fig. 6.1 Cultivation practice in Khatijapura

The farmers were also enquired about the general rates for different practices in vogue. The rates as informed by the farmers are as follows:

• Contract farming: rate per acre ranges between Rs. 7000/− and Rs. 9000/−.
• Crop sharing: harvested crop is shared in the ratio of 2:1 between the owner and cultivator, respectively.
• Leased out: not practiced much.

6.1.3 Cropping Pattern

Rabi season starts from end of September or early October, and the harvest continues up to January (this year it extended up to mid-February). *Rabi* crops observed in Khatijapura are given in Table 6.1 (Fig. 6.2 and Picture 6.1).

Farmers generally practice seed drilling method for sowing, and a few have been using machineries for harvesting. Chemical fertilizers such as DAP and urea are being applied (20 to 25 kgs/acre) in most of the crop plots. Insecticides such as

Table 6.1 *Rabi* crops observed in Khatijapura

Crop pattern	Crop category	Crop name	Variety
Single crop	Cereals	*Jowar* (Sorghum)	Local
		Maize	Local and hybrid
		Wheat	Local
	Flowers	Chrysanthemum	Local and hybrid
	Fodder crops	Fodder *Jowar*	Local
		Napier grass	Local
	Oil seeds	Safflower	Local
	Pulses	Bengal gram	Local
		Black gram	Local
Single crop	Vegetable crops	Cucumber	Local
		Chili	Local
		Brinjal	Local
		Beans	Local
Mixed crop	Pulses + cereals	Bengal gram + *Jowar* (Sorghum)	Local
	Cereals	Wheat + *Jowar* (Sorghum)	Local
Annual	Commercial crop	Sugarcane	Local and hybrid
Perennial	Fruit crops	Pomegranate	Hybrid
		Lemon	Local
		Guava	Local and hybrid
	Commercial crop	Mulberry	Local

Source: Field inventory during February 2017 by PIXEL team

Khatijapura – Rabi Season (2016-17)

Crop	Extent in Ha.
Black gram	2.99
Chilly	0.12
Beans	0.34
Bengalgram	97.21
Bengalgram + Jowar	7.99
Brinjal	1.06
Chrysanthemum	0.38
Cucumber	0.26
Jowar	125.66
Maize	2.91
Safflower	10.13
Wheat	42.15
Wheat + Jowar	1.29
Fodder Jowar	0.25
Napier grass	0.27
Total	**293.01**

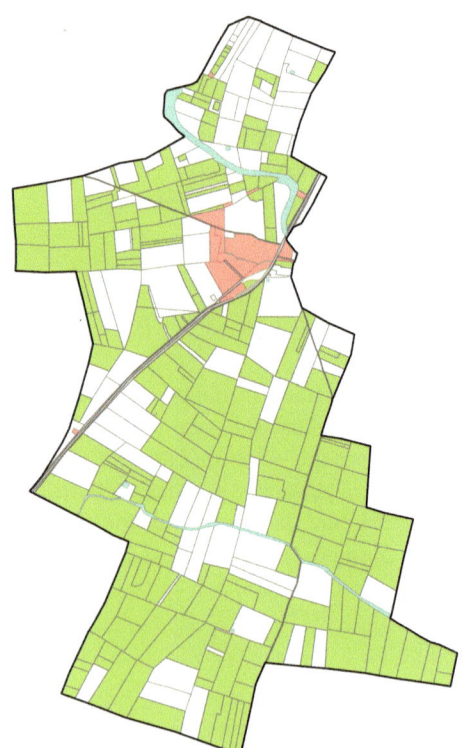

Crop categories - Rabi season, Khatijapura

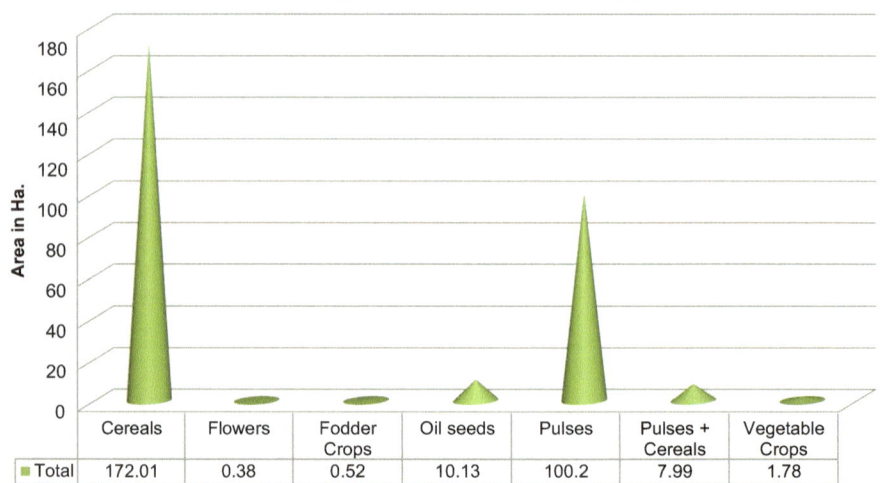

	Cereals	Flowers	Fodder Crops	Oil seeds	Pulses	Pulses + Cereals	Vegetable Crops
Total	172.01	0.38	0.52	10.13	100.2	7.99	1.78

Fig. 6.2 Extent of crops in Khatijapura during *Rabi* season (2016–2017)

Picture 6.1   Crops inventoried in Khatijapura (*Rabi* 2016–2017)

Daman 47, Proton, Neutron, Super Nova, etc. are being used specifically for tur, safflower, and Bengal gram crops. One of the famers has been found to be practicing organic farming.

6.1.4   Summer Crops

Cultivation during summer in Khatijapura appears to be controlled by the developed groundwater-based irrigation sources. Very few holdings have crops during the summer, and the total cropped area is 3.03 Ha (Fig. 6.3 and Picture 6.2).

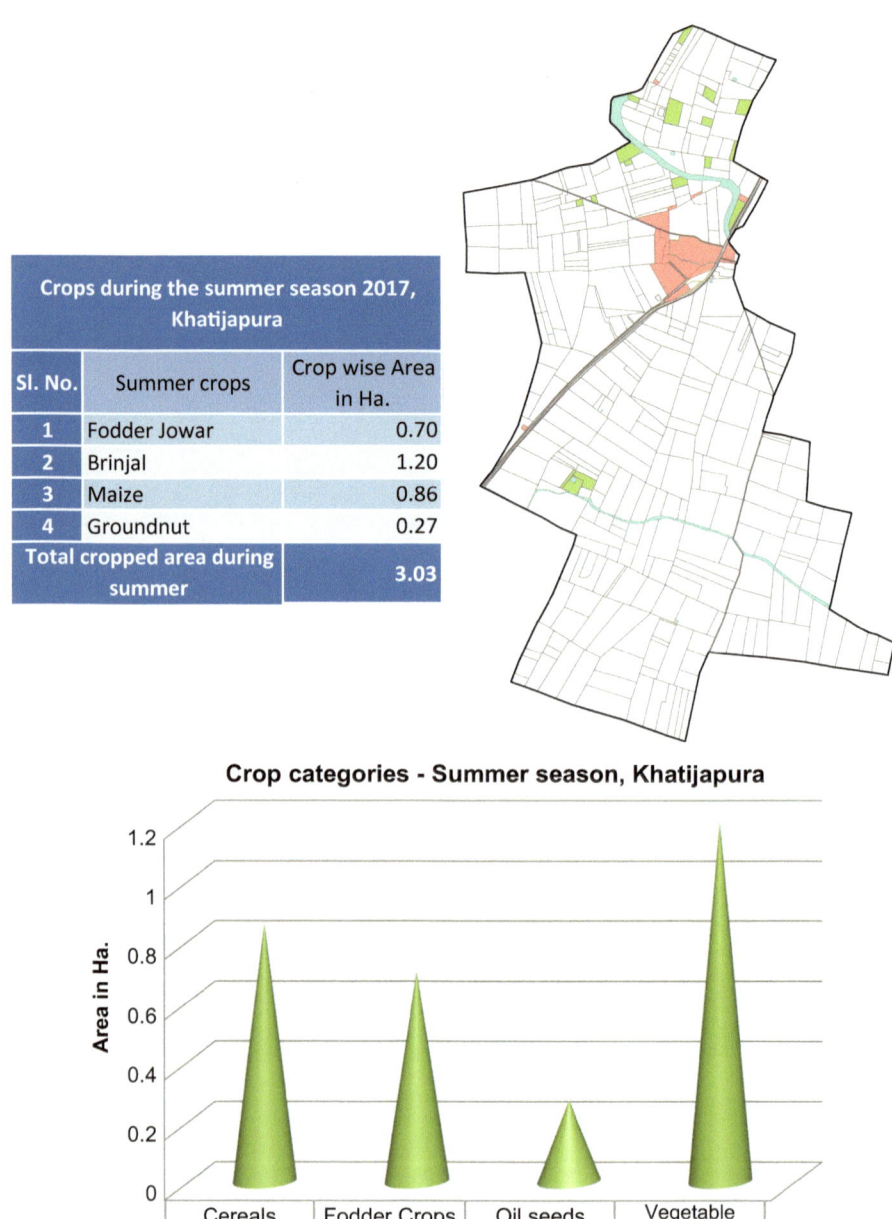

Crops during the summer season 2017, Khatijapura		
Sl. No.	Summer crops	Crop wise Area in Ha.
1	Fodder Jowar	0.70
2	Brinjal	1.20
3	Maize	0.86
4	Groundnut	0.27
Total cropped area during summer		3.03

Fig. 6.3 Spatial spread of summer crops in Khatijapura

Picture 6.2 Summer crops in Khatijapura

Detailed plot-wise inventory in Khatijapura is provided below. The village has 324 crop plots. During the summer 2017, famers have grown crops only in nine plots (Table 6.2) covering an area of 3.03 Ha.

It was also observed that only a portion of the entire plot has been cultivated, mostly based on the availability of the irrigation source. Further, 1.64 ha of land has been cultivated both in *Rabi* and summer (Tables 6.3 and 6.4).

6.1.5   **Kharif** *Season (August 2017)*

During inventory 350 plots were inventoried. Out of which 136 plots had standing *Kharif* crops. Crops enumerated in *Kharif* season (Table 6.5) are as follows (Pictures 6.3 and 6.4).

Annual and perennial types of crops are noticed in 9 plots having cumulative area of 4.89 Ha., and details are as follows (Table 6.6).

In Khatijapura there is a distinct cropping pattern, and crops in *Rabi* and *Kharif* are different. Farmers generally go for pulses during *Kharif* and cereals during *Rabi*. Cropping pattern during all the three seasons are provided in the Table 6.7 .

6.1.6   *Plots Cultivated in Different Seasons*

The agricultural plots cultivated in different seasons (Table 6.8) are presented below.

Table 6.2 Crops during the summer season 2017, Khatijapura

Sl. No.	Summer crops	Survey No.	Plot ID	Plot area (in Ha.)	Farmer name	Plot-wise crop area (in Ha.)	Crop wise area in Ha
1	Fodder Jowar	68/1b	315	0.58	Sharanappa Dhareppa Shyapeti	0.58	0.7
2		61/1	327	0.12	Siddanagowda Shivangowda Patil	0.12	
3	Brinjal	62/2	14	0.39	Laxman Govindappa Valikar	0.39	1.2
5		47/1	112	0.64	Laxman Siddappa Gavali	0.64	
6		47/1	107A	1.81	Laxman Siddappa Gavali	0.17	
7	Maize	66/1	211A	1.42	Khairunnishabeg Usman Nayak	0.59	0.86
4		64/2	203A	1.17	Laxman Govindappa Valikar	0.27	
8	Groundnut	58/2	48A	1.98	Ramappa Amogi Gavali	0.13	0.27
9		58/1	319A	0.45	Muttappa Amogi Gavali	0.14	
Total cropped area during summer						3.03	3.03

Table 6.3 Area cultivated in both *Rabi* and *Summer* seasons

Plot ID	Plot area (in Ha)	Season				Gross cropped area during *Rabi* and summer (in Ha)
		Rabi	Area (in Ha)	Summer	Area (in Ha)	
48	1.91	Black gram	1.91	Groundnut	0.13	2.04
107	1.81	Wheat	1.81	Brinjal	0.17	1.98
112	0.64	Brinjal	0.64	Brinjal	0.64	1.28
315	0.58	Wheat	0.58	Fodder Jowar	0.58	1.16
327	0.12	Chili	0.12	Fodder Jowar	0.12	0.24
Total	5.06	*Rabi* total	5.06	*Summer* total	1.64	6.7

Table 6.4 Crops statistics as enumerated during *Rabi* and *Summer* seasons

	Rabi		Summer		Annual	Perennial	
Crop	Rainfed	Irrigated	Rainfed	Irrigated	Irrigated	Irrigated	Total
Black gram	2.99						2.99
Chili		0.12					0.12
Beans	0.34						0.34
Bengal gram	96.57	0.64					97.21
Bengal gram + *Jowar*							7.99
Brinjal	0.42	0.64		1.20			2.26
Chrysanthemum	0.26	0.12					0.38
Cucumber	0.26						0.26
Jowar	125.66						125.66
Maize	1.33	1.58		0.86			3.77
Safflower	10.13						10.13
Wheat	33.65	8.50					42.15
Wheat + *Jowar*	0.31	0.98					1.29
Fodder *Jowar*		0.25		0.70			0.95
Napier grass		0.27					0.27
Sugarcane					1.58		1.58
Guava						0.88	0.88
Lemon						0.35	0.35
Mulberry						0.43	0.43
Pomegranate						0.93	0.93
Groundnut				0.27			0.27
Total	279.91	13.10	0.00	3.03	1.58	2.59	300.21

The table is based on the crops inferred for *Kharif* 2016–2017 as the inventory started in *Rabi* only. Due to rainfall availability and the convenience of cultivation, farmers tend to cultivate different plots in different season. During *Kharif* (Sept 2017), due to delayed monsoon, a number of plots have been reduced (Picture 6.5). But farmers feel that they may be able to grow in more plots as rainfall has improved later. This was observed during the *Kharif* inventory. Many plots were kept ready for sowing during *Rabi* (2017–2018). It is also observed that a few plots cultivated during *Rabi* (2016–2017) have been cultivated during *Kharif* (Sept 2017) (Picture 6.6).

6.2 Daliparru

Daliparru village is a medium-sized village covering 581.23 ha of land. It is a part of Krishna-Godavari Zone (agroclimatic zone) – coastal plains of Andhra Pradesh and the agriculture is supported by an irrigation project. The village is characterized

Table 6.5 Crops during *Kharif* season (2017–2018)

Crop pattern	Crop category	Crop name	No. of plots	Area in Ha.
Single crop	Cereals	*Jowar*	5	1.89
		Maize	14	8.37
	Pulses	Black gram	6	14.62
		Tur	70	103.92
	Vegetable crops	Brinjal	1	0.20
		Cucumber	8	3.76
		Tomato	2	1.39
		Lady's finger	2	0.42
		Coriander	1	0.32
	Tubers	Onion	9	6.17
	Flowers	Gerbera	2	0.35
		Chrysanthemum	1	0.05
	Fodder crops	Fodder *Jowar*	2	0.39
Sub Total			123	141.85
Mixed crops	Pulses	Tur + black gram	3	0.78
	Pulses + cereals	Tur + *Jowar*	4	6.77
	Pulses + vegetable crops	Tur + cucumber	1	0.47
		Tur + lady's finger	1	0.51
	Vegetable crops	Coriander + leafy vegetables	1	0.11
		Coriander + cluster bean	1	0.13
		Cucumber + coriander	1	0.21
	Vegetable crops + tubers	Brinjal + ridge gourd + onion	1	0.32
Subtotal			13	9.31
Total			136	151.17

by small to medium holdings with few exceptions, and the plots are fairly large. About 1182 crop plots (including plantation) are inventoried with minimum area of 0.66 Gunta and maximum of 5 acres 09 Gunta. Average area of crop plot in this village is 1 acre 2 Gunta.

6.2.1   Sources of Irrigation

Daliparru village is irrigated by the canal water from Prakasam Barrage (Vijayawada) Krishna River. No irrigation borewells and open wells are noticed in this village because groundwater is brackish to saline and not useful for irrigation. There are three pump houses to lift the water from the pond (water stored from stream and excess canal water).

Picture 6.3 Geographical
spread of *Kharif* crop plots
in Khatijapura

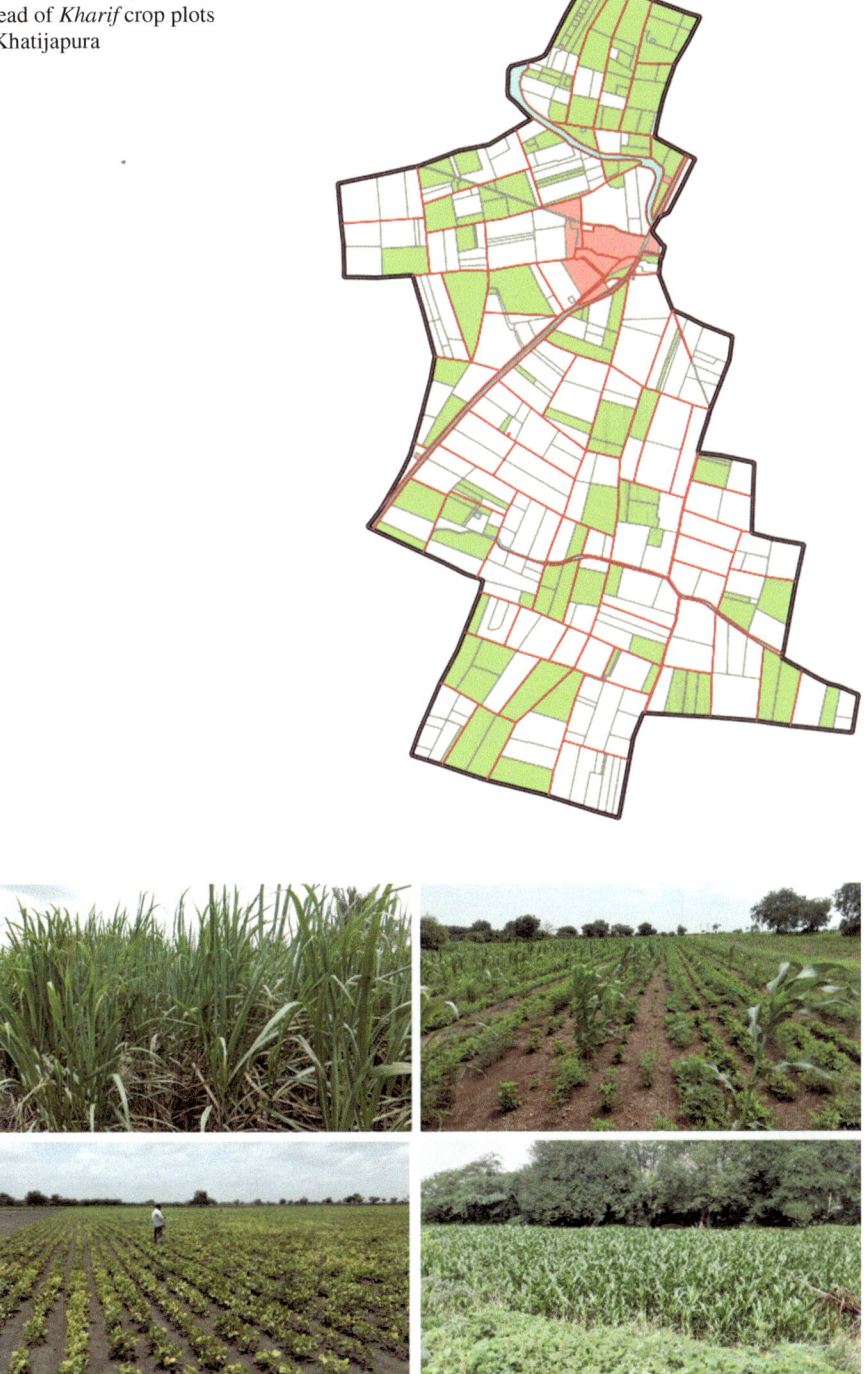

Picture 6.4 *Kharif* Crops in Khatijapura (August 2017)

Table 6.6 Details of plots having cumulative area

Crop category	Crop name	No. of plots	Area in Ha.
Annual	Sugarcane	4	2.3
Perennial	Guava	2	0.88
	Pomegranate	1	0.93
	Mulberry	1	0.43
	Lemon	1	0.35
Total			4.89

6.2.2   Cultivation Practices

The village is located 18 kms from Machalipatnam Town and 5 kms from Ghantasala Mandal headquarters; about 15% of land holders have migrated to cities, and about 7% of land holdings belong to NRI's. Therefore, lease cultivation (15%) practice is noticed in this village.

1. Leased out (leasing rate): Rate per acre ranges between Rs. 18,000/− and Rs. 22,000/−.

6.2.3   Cropping Pattern

In Daliparru, the *Rabi* season starts from end of November or mid of December, and the harvest continues up to March. *Rabi* crops noticed in Daliparru are presented in Table 6.9.

Extent of crops inventoried using GPS during *Rabi* season (Fig. 6.4) are as follows (Picture 6.7).

In Daliparru, farmers generally practice broadcasting method for sowing, and all the farmers adopt only manual harvesting methods. Chemical fertilizers such as DAP and urea are being applied (20 to 25 kgs/acre) in most of the crop plots in *Kharif* season, but in *Rabi* season, urea is very rarely used depending upon water availability. Pesticides such as uthane M45, monocrotophos, Cloro20, Saaf, Nativa, Revan, Coragen, Ergon, and Dhanzyme Gold are applied depending upon the type of disease.

6.2.4   Summer Crops

Daliparru village has 1182 crop plots (including plantation area). During the summer 2017, famers have grown crops only in two plots covering an area of 0.73 Ha (Pictures 6.8 and 6.9; Table 6.10).

Table 6.7 Overall cropping pattern – Khatijapura

Pattern	Category	Crop	Area in Ha.							Total	
			Rabi (Feb 2017)		Summer (Apr 2017)	Kharif (August 2017)		Annual	Perennial		
			Rainfed	Irrigated	Irrigated	Rainfed	Irrigated	Irrigated	Irrigated		
Single crop	Cereals	Jowar	125.66			1.89				127.55	181.85
		Maize	1.33	1.58	0.86	5.32	3.06			12.14	
		Wheat	33.65	8.50						42.15	
	Pulses	Black gram	2.99			10.25	4.37			17.61	218.74
		Tur				103.92				103.92	
		Bengal gram	96.57	0.64						97.21	
	Oil seeds	Safflower	10.13							10.13	10.4
		Groundnut			0.27					0.27	
	Tubers	Onion				5.76	0.41			6.17	6.17
	Vegetables	Beans	0.34							0.34	9.07
		Brinjal	0.42	0.64	1.20	0.20				2.46	
		Cucumber	0.26			2.51	1.24			4.02	
		Coriander				0.32				0.32	
		Lady's finger				0.42				0.42	
		Tomato				1.39				1.39	
		Chili		0.12						0.12	
	Flowers	Chrysanthemum	0.26	0.12			0.05			0.43	0.78
		Gerbera					0.35			0.35	
	Fodder crops	Fodder Jowar		0.25	0.70		0.39			1.34	1.61
		Napier grass		0.27						0.27	

(continued)

Table 6.7 (continued)

		Area in Ha.								
Mixed crop	Tur + *Jowar*				6.77				6.77	18.59
	Tur + safflower								0.00	
	Tur + black gram				0.78				0.78	
	Bengal gram + *Jowar*	7.99							7.99	
	Wheat + *Jowar*	0.31	0.98						1.29	
	Tur + cucumber					0.47			0.47	
	Tur + lady's finger					0.51			0.51	
	Brinjal + onion + ridge gourd					0.32			0.32	
	Coriander + cluster bean					0.13			0.13	
	Coriander + leafy vegetables				0.11				0.11	
	Cucumber + coriander				0.21				0.21	
Annual/fruit crops	Sugarcane						2.30		2.30	4.89
	Guava							0.88	0.88	
	Lemon							0.35	0.35	
	Mulberry							0.43	0.43	
	Pomegranate							0.93	0.93	
	Total	279.91	13.1	3.03	139.85	11.3	2.3	2.59	452.08	

Table 6.8   Plots cultivated during the year 2016–2017 in village of Khatijapura

Type	No of plots	Total area in Ha.	Crops grown	Reasons
All season plots	9	4.89	Sugarcane, guava, pomegranate, mulberry, lemon	Crops are irrigated by ground water resource
Two season plots	9	8.49	Tur + brinjal, tur + maize, tur + groundnut, black gram + groundnut, wheat + brinjal, wheat + fodder *Jowar*, chili + fodder *Jowar*	Crops are irrigated by groundwater resource
Single season (plots during *Kharif* (2016–2017) are inferred)	323	430.86	Beans, Bengal gram, black gram, brinjal, chili, chrysanthemum, cucumber, fodder *Jowar*, *Jowar*, maize, napier grass, onion, safflower, sunflower, tur, wheat	Dependent on rain and soil moisture
Single season (plots during *Kharif* (Sept 2017 as enumerated)	274	348.13	Tur, *Jowar*, onion, maize, wheat, chrysanthemum, Bengal gram, safflower, fodder *Jowar*, black gram, cucumber, coriander, brinjal, ridge gourd, cluster bean, napier grass, lady's finger, chili, beans	Due to delayed monsoon, farmers have cultivated reduced plots

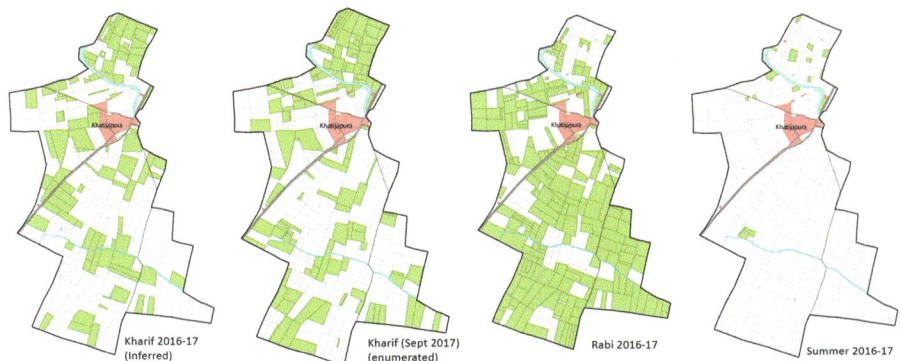

Kharif 2016-17 (Inferred)   Kharif (Sept 2017) (enumerated)   Rabi 2016-17   Summer 2016-17

Picture 6.5   Plots cultivated in different seasons, Khatijapura village

Two plots that had mango plantations earlier have been removed, and the land is prepared for paddy. The summer crop (Fodder *Jowar*) is supported by the waste water from the village, and the Gini grass is supported by water from an open well (adjacent to poultry farm). Both the plots having crops in the summer had Black gram during the *Rabi* season (Table 6.11).

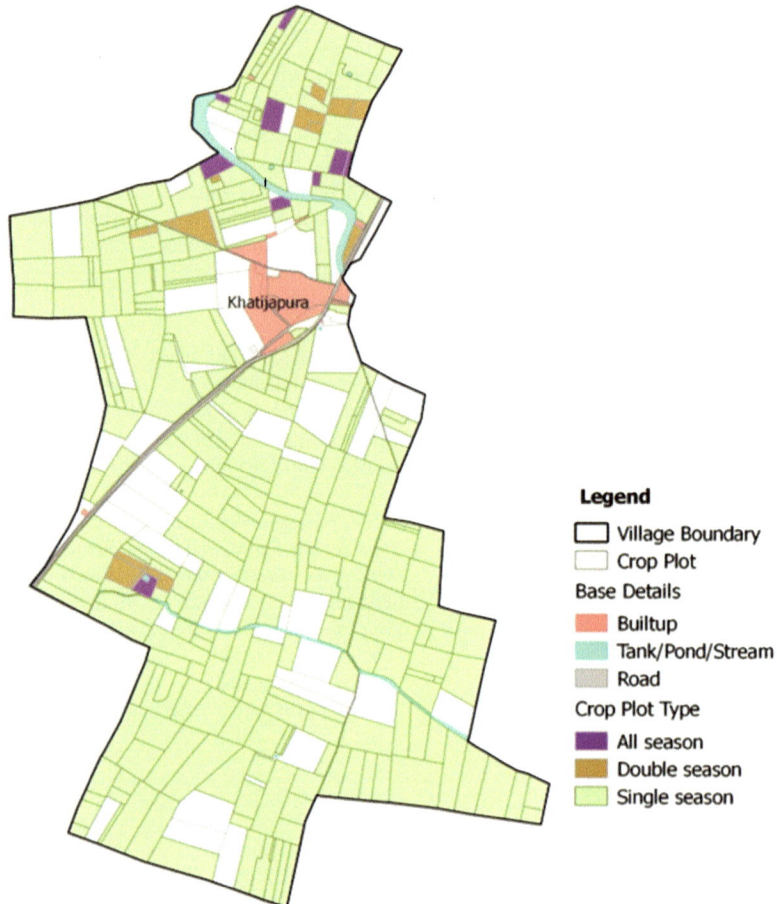

Picture 6.6 Crop plots of all seasons in Khatijapura

Table 6.9 *Rabi* crops noticed in Daliparru

Crop pattern	Crop category	Crop name	Variety
Single crop	Pulses	Black gram	Local
		Horse gram	Local
	Fodder crops	Fodder *Jowar*	Local
		Gini grass	Local
		Sunn hemp	Local
		Others (Saijulu)	Local
	Tubers	Turmeric	Local
Perennial	Plantation and spice crops	Coconut	Local
Perennial	Fruit crops	Mango	Local

| Daliparru - Rabi (2016-17) ||
Crop	Area in Ha.
Black Gram	485.71
Green Gram	0.00
Fodder Jowar	0.15
Gini Grass	0.39
Horse Gram	0.15
Others (Saijulu)	0.17
Sunn hemp	3.36
Turmeric	0.11
Total	490.04

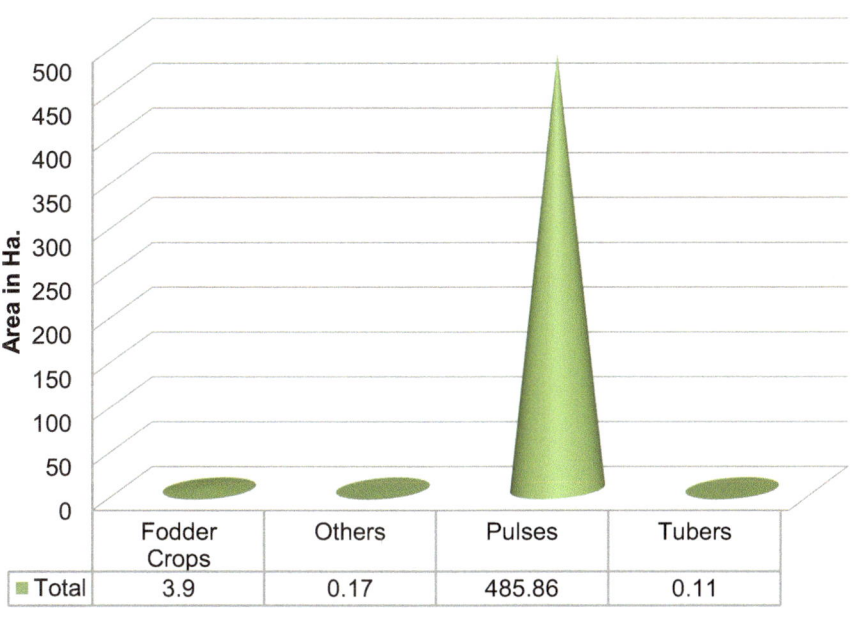

Crop categories - Rabi season, Daliparru

	Fodder Crops	Others	Pulses	Tubers
■ Total	3.9	0.17	485.86	0.11

Fig. 6.4 Extent of crop inventoried in *Rabi* season

6.2.5 Kharif *Crops*

Climatic conditions in Daliparru are entirely different from Khatijapura, and the cropping is also more of paddy. Daliparru village has 1182 crop plots (including plantation area). During the *Kharif* (August 2017), almost all the plots (1152) had paddy crop (Table 6.12) covering an area of 499.44 Ha (Pictures 6.10 and 6.11; Table 6.13).

Picture 6.7   Crops inventoried during *Rabi* season (2016–2017) in Daliparru

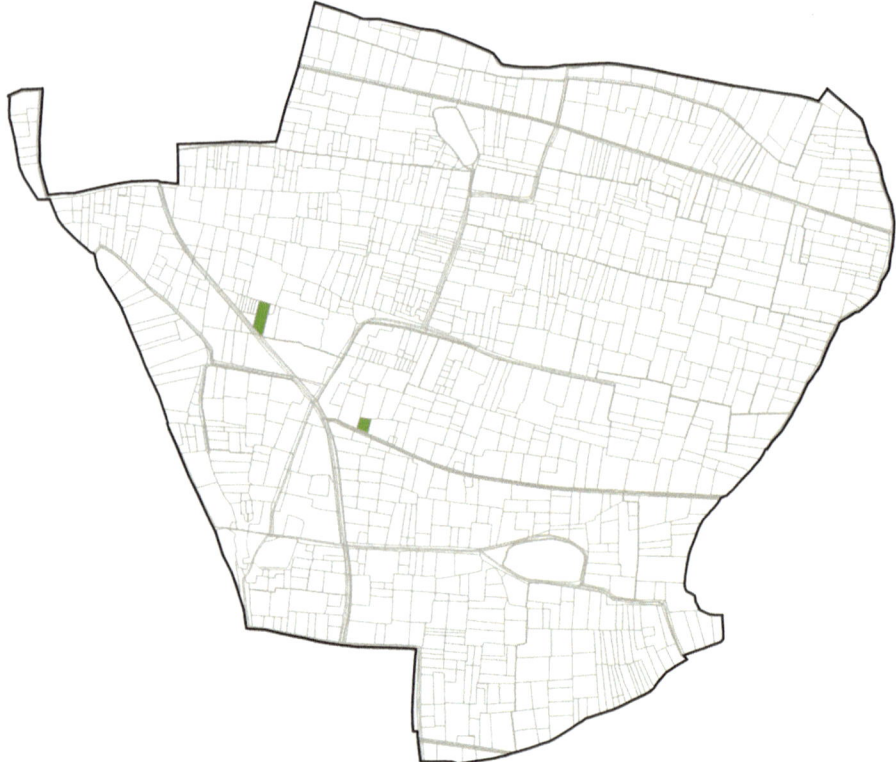

Picture 6.8   Spatial spread of Summer crops in Daliparru

Picture 6.9 Summer crops in Daliparru

Table 6.10 Crops during the Summer season 2017, Daliparru

Sl. No.	Summer crops	Survey no.	Plot ID	Area (in Ha.)	Farmer name
1	Gini grass	135/2135/1	1420	0.23	Shivakoteshwar Rao
2	Fodder *Jowar*	45–3	1316	0.51	G Srinivasa Rao
Total				0.73	
3	Earlier plantation removed	284/3285/4	1418B	0.03	V Lalitha
4		135/3	1595A	0.06	V Lalitha
Total				0.09	

Table 6.11 Crops statistics as enumerated during *Rabi* and *Summer* seasons

Crop	*Rabi*		Summer		Annual	Perennial	Total in Ha.
	Rainfed	Irrigated	Rainfed	Irrigated	Rainfed	Rainfed	
Black gram	485.71						485.71
Coconut						0.25	0.25
Fodder *Jowar*	0.15			0.51			0.66
Gini grass	0.39			0.23			0.62
Horse gram	0.15						0.15
Mango						0.56	0.56
Others (Saijulu)	0.17						0.17
Sunn hemp	3.36						3.36
Turmeric		0.11					0.11
Total	489.93	0.11		0.73	0.00	0.81	491.58

Table 6.12 *Kharif* crops in Daliparru Village (September 2017)

Crop	*Kharif*		Perennial (Rainfed)	Total (area in Ha)
	Rainfed	Irrigated		
Paddy		499.44		499.44
Coconut			0.25	0.25
Fodder *Jowar*	1.63	0.82		2.45
Mango			0.56	0.56
Lady's finger	0.17			0.17
Total	1.80	500.26	0.81	502.86

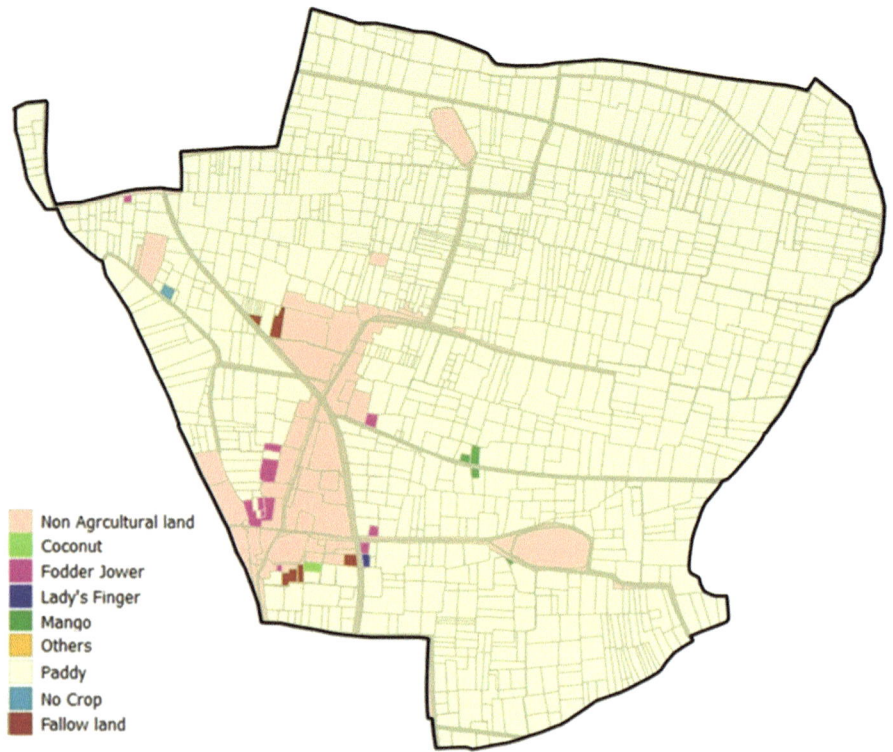

Non Agrcultural land
Coconut
Fodder Jower
Lady's Finger
Mango
Others
Paddy
No Crop
Fallow land

Picture 6.10   Distribution of paddy crops in Daliparru (September 2017)

Picture 6.11   *Kharif* crops in Daliparru (September 2017)

Table 6.13 Cropping pattern in Daliparru

Crop pattern	Crop category	Crop name	Rabi (Feb 2017) Area in Ha	Summer (May 2017) Area in Ha	Kharif (Sept 2017) Area in Ha
Single crop	Cereals	Paddy			499.44
	Pulses	Black gram	486.02		
		Horse gram	0.15		
	Fodder crops	Fodder *Jowar*	0.15	0.51	2.45
		Gini grass	0.39	0.23	
		Others (Saijulu)	0.17		
		Sunn hemp	3.36		
	Tubers	Turmeric	0.11		
	Vegetable crops	Lady's finger			0.17
	Perennial	Coconut	0.25	0.25	0.25
		Mango	0.65	0.56	0.56
Total			491.25	1.54	502.86

6.2.6   Plots Cultivated in Different Seasons

In the village Daliparru, *Kharif* crops are grown with canal water, and the residual soil moisture supports the *Rabi* crops. As observed, the village has pattern of cultivation of almost all the plots except a few during both *Kharif* and *Rabi* seasons and mono-cropping prevails (*Kharif*, paddy and *Rabi*, pulses: black gram). During the *Rabi* season inventory, the remnants of earlier paddy were clearly visible, and hence inference was drawn. The inference was corroborated during the *Kharif* (Sept 2017) inventory that paddy occupied all the plots (Picture 6.12; Table 6.14).

6.3   Jarasingha

Jarasingha is a small village covering 367.86 ha of land. It is a part of mid-central table land (agroclimatic zone) interior coastal midland of Odisha, and the agriculture is supported by an irrigation project. The village is characterized by very small to small holdings. About 4508 crop plots (including plantation) are inventoried with minimum area of 0.2 Gunta and maximum of 5 acres 30 Guntas. Average area of crop plot in this village is 5 Guntas.

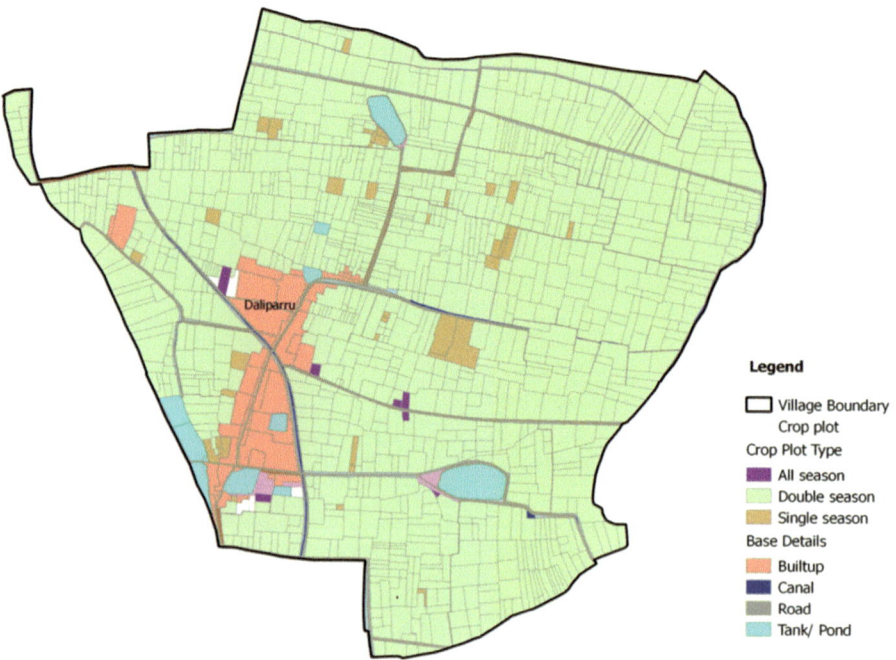

Picture 6.12 Crop plots of all seasons in Daliparru

Table 6.14 Plots cultivated during the year 2016–2017 in the village of Daliparru

Type	No. of plots	Total area in Ha.	Crops grown	Reasons
All season plots	7	1.54	Paddy, black gram, fodder Jower, gini gross, mango, coconut	Cultivated as source of water is available
Two season plots	1123	489.45	Paddy, black gram, horse gram, sunn hemp, gini grass, fodder Jower, lady's finger, others (Saijulu), turmeric, and mango	*Kharif* season crops are cultivated with canal water, and during *Rabi* the residual soil moisture supports crops
Single season plots	40	12.12	Paddy, black gram, fodder *Jowar*, and turmeric	Few farmers have their convenience of cultivation either during *Kharif* or *Rabi*

6.3.1   Sources of Irrigation

Jarasingha villages come under the Derjang Reservoir canal command area, and hence surface water irrigation is practiced. Few open wells and borewells are noticed in the surrounding of settlement where people use the groundwater for cultivating vegetable crops throughout the year.

- There are 26 open wells noticed, and general depth of open wells ranges from 25 to 35 feet; diameter of well ranges from 5 to 12 feet, and water level from the BGL is 20 to 25 feet.
- There are 7 borewell noticed, and general depth of borewells ranges from 250 to 400 feet yield ranges from 1.5 to 2.5″.
- Dried open wells are noticed in northern side of the village (New Jarasingha locality).

6.3.2 Cultivation Practices

Jarasingha village has two settlements – New Jarasingha on SH-63 and Old Jarasingha nearby river. Since, the village is located 5 kms from Angul City, almost all the big land owners reside in Angul city and practice contract farming, crop sharing, and leasing out of farm lands. In general, the crop sharing method are practiced in canal-irrigated area (paddy), whereas leased out method are practiced in groundwater-irrigated area (vegetable crops).

- Crop sharing: Harvested crop will be shared in the ratio of 50:50 (including expenditure) between the owner and cultivator, respectively.
- Leased in: 1500 to 2000 /– per Gunta/year.

6.3.3 Cropping Pattern

Rabi season starts from November, and the harvest continues up to end of February. The crops observed and enumerated (Table 6.15) are as follows.

Extent of crops inventoried using GPS during *Rabi* season (Fig. 6.5) are as follows (Picture 6.13).

In Jarasingha farmers generally practice seed drilling and broadcasting method for sowing, and all the farmers adopt only manual harvesting methods.

Chemical fertilizers such as growmore, potash/DAP, gypsum, and urea are being applied (20 to 25 kgs/acre) in most of the crop plots. Insecticides such as hormone are used specifically for vegetable crops (lady's finger etc.).

6.3.4 Summer Crops

Jarasingha village had 4508 crop plots as per the inventory during *Rabi* season. During the summer 2017, crops have been observed only in 23 plots covering an area of 0.71 Ha. During the *Rabi* season, presence of Kitchen Garden within the Village was observed. Since such features were within the settlement, they were not separately mapped, and the area was considered as built up.

Table 6.15 *Rabi* crops noticed in Jarasingha

Crop pattern	Crop category	Crop name	Variety
Single crop	Fodder crops	Fodder chani	Local
	Oil seeds	Groundnut	Local
	Pulses	Green gram	Local
		Black gram	Local
	Vegetable crops	Cucumber	Local
		Chili	Local
		Brinjal	Local
		Beans	Local
		Green peas	Hybrid + local
	Leafy vegetables	Dantu	Hybrid
Mixed crop	Leafy vegetables+ tubers	Dantu (Amaranthus) + coriander + onion + garlic	Hybrid + local
	Pulses + vegetable crop	Green gram + green peas	Local + hybrid
	Pulses + vegetable crop	Black gram + green peas	Local + hybrid
	Leafy vegetables + tubers	Lady's finger + Dantu + onion + garlic	Hybrid + local
Perennial	Fruit crops	Mango	Hybrid + local

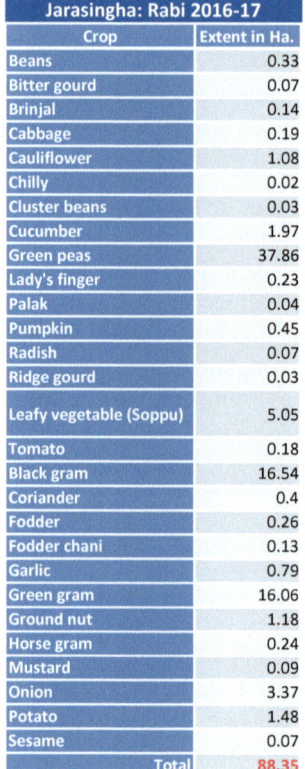

Jarasingha: Rabi 2016-17	
Crop	Extent in Ha.
Beans	0.33
Bitter gourd	0.07
Brinjal	0.14
Cabbage	0.19
Cauliflower	1.08
Chilly	0.02
Cluster beans	0.03
Cucumber	1.97
Green peas	37.86
Lady's finger	0.23
Palak	0.04
Pumpkin	0.45
Radish	0.07
Ridge gourd	0.03
Leafy vegetable (Soppu)	5.05
Tomato	0.18
Black gram	16.54
Coriander	0.4
Fodder	0.26
Fodder chani	0.13
Garlic	0.79
Green gram	16.06
Ground nut	1.18
Horse gram	0.24
Mustard	0.09
Onion	3.37
Potato	1.48
Sesame	0.07
Total	88.35

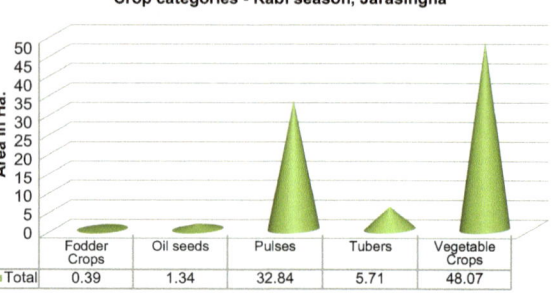

Crop categories - Rabi season, Jarasingha

	Fodder Crops	Oil seeds	Pulses	Tubers	Vegetable Crops
Total	0.39	1.34	32.84	5.71	48.07

Fig. 6.5 Extent of crops inventoried in *Rabi* season

Picture 6.13 Crops enumerated during *Rabi* season (2016–2017) in Jarasingha

During summer also the same pattern was observed. Generally, each household has 10x20m or slightly larger plots and portion of which is built, and the rest is used for growing different types of vegetables that has influence on the livelihood of the farmers. The produces from this garden are sold locally and sent to Angul city also. Therefore, an attempt to map the area of Kitchen Garden was made during summer using GPS. These Kitchen Gardens have been captured as small plots. Sixty-four plots having Kitchen Garden could be mapped, and cumulative area of such garden crops within the settlement is 1.53 Ha. However, the house plots having very small Kitchen Garden could not be delineated and therefore included in the settlement area itself. Totally the village had 2.25 Ha. of crop area during the summer 2017 (Table 6.16).

There was constraint in getting the farmers name linked to the crop plot. The farmers were not cooperating, and correlating the revenue records with the map generated using GIS was difficult (Table 6.17). However, during the summer inventory (Picture 6.14), few plots could be correlated/identified with farmer. Wherever such correlation was possible, it was recorded (Picture 6.15; Table 6.18).

6.3.5   Kharif *Season*

Jarasingha presents a conspicuous pattern of cultivation. The landholdings are very small and are having irrigation facility also. In the earlier inventory, minor crop plots in the periphery and those could not be properly ascertained were included in

Table 6.16 Crops during the Summer season 2017, Jarasingha

Sl. No.	Crop	Area in Ha.
1	Beans	0.06
2	Bitter guard	0.07
3	Brinjal	0.02
4	Chili	0.02
5	Lady's finger	0.52
6	Pumpkin	0.18
7	Pudina	0.03
8	Ridge gourd	0.15
9	Snake gourd	0.05
10	Spinach	0.12
11	Til (Sesamum)	0.11
12	Leafy vegetable (Soppu)	0.91
13	Mixed crop(cucumber+ ridge gourd)	0.01
	Total	2.25

the settlement, as built up. Few mixed plantation boundaries could not be accurately traced as they did not have neat boundary. Having observed Kitchen Garden all across the settlement, one more careful interpretation of the image (December 2016) was done, and the settlement boundary was appropriately delineated. Further this helped to map the tiny Kitchen Garden also, and obscured boundaries of mixed plantation could be rectified. This resulted in a good GIS map for field inventory.

Though paddy cultivation dominates, vegetables constitute substantial portion of the cropping. During *Kharif* (September 2017) 3893 plots were inventoried with different crops. Cereals occupy 74.5% of the area cultivated, while the vegetables (different varieties) cover 7.2% of the cultivated area and pulses of different types have a share of 5.4%. Among cereals, paddy itself has occupied an area of 141.61 Ha (99.5%). The details are presented in the table below (Pictures 6.16 and 6.17; Tables 6.19 and 6.20).

6.4   Plots Cultivated in Different Seasons

In the village of Jarasingha, *Kharif* crops are grown with canal water, and the residual soil moisture supports the *Rabi* crops. As observed, the village has dominance of paddy during *Kharif* with vegetables and little pulses. The *Rabi* has more of pulses supported by residual soil moisture and vegetables supported by groundwater. The remnants of *Kharif* crops could not be observed in all the plots during the *Rabi* inventory, and therefore pattern of cultivation of plots is arrived based on *Rabi* (2016–2017, Summer 2016–2017, and *Kharif* 2017–2018) (Picture 6.18; Table 6.21).

Table 6.17 Crops statistics as enumerated during *Rabi* and *Summer* seasons

	Rabi		Summer		Annual	Perennial		Total
Crop	Rainfed	Irrigated	Rainfed	Irrigated	Irrigated	Rainfed	Irrigated	area in Ha.
Banana					0.16			0.16
Beans		0.33		0.06				0.39
Bitter guard		0.07		0.07				0.14
Black gram	16.54							16.54
Brinjal		0.14		0.02				0.16
Cabbage		0.19						0.19
Cauliflower		1.08						1.08
Chili		0.02		0.02				0.04
Cluster beans		0.03						0.03
Coconut							0.39	0.39
Coriander		0.40						0.40
Cucumber		1.97						1.97
Fodder	0.26							0.26
Fodder *chani*	0.13							0.13
Garlic		0.79						0.79
Green gram	16.06							16.06
Green peas	37.86							37.86
Groundnut		1.18						1.18
Horse gram	0.24							0.24
Lady's finger		0.23		0.52				0.74
Mango plantation						12.55	0.70	13.26
Mixed plantation (Mango, teak, bamboo & others trees)						5.97		5.97
Mustered	0.09							0.09
Onion		3.37						3.37
Palak		0.04						0.04
Potato		1.48						1.48
Pumpkin		0.45		0.14				0.59
Pudina				0.03				0.03
Radish		0.07						0.07
Ridge gourd		0.03		0.15				0.18
Snake gourd				0.05				0.05
Spinach				0.12				0.12
Sesamum-til	0.07			0.11				0.18
Leafy vegetable (*Soppu*)		5.05		0.95				6.00
Tomato		0.18						0.18

(continued)

Table 6.17 (continued)

Crop	Rabi		Summer		Annual	Perennial		Total area in Ha.
	Rainfed	Irrigated	Rainfed	Irrigated	Irrigated	Rainfed	Irrigated	
Tur								0.00
Mixed crop(cucumber+ ridge gourd)				0.01				0.01
Total	71.24	17.10	0.00	2.25	0.16	18.52	1.09	110.32

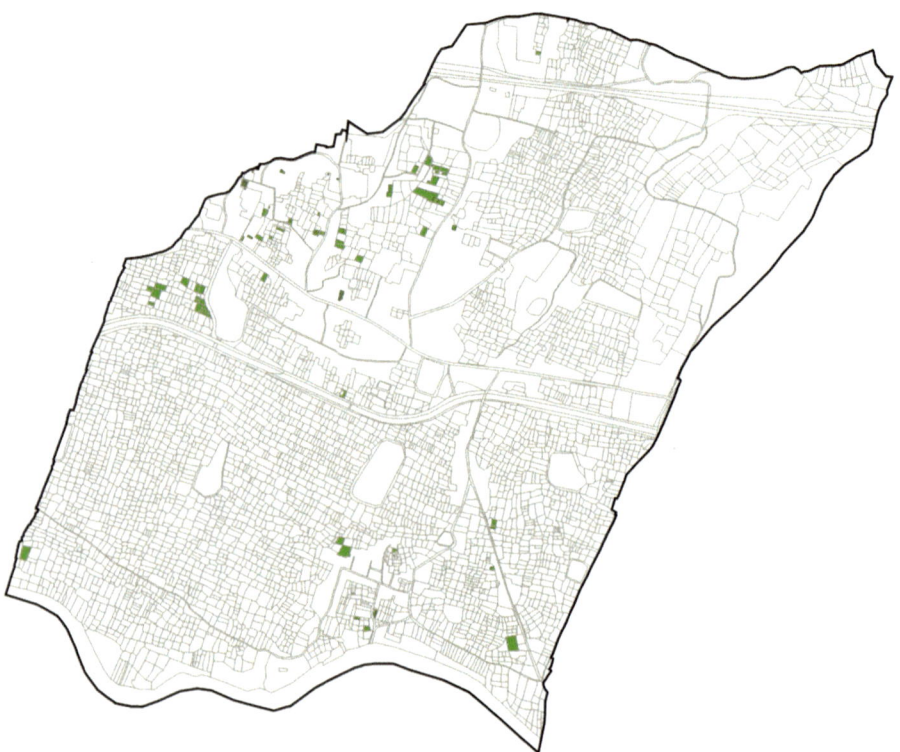

Picture 6.14 Disposition of crop plots during the Summer 2017, Jarasingha

Picture 6.15   Kitchen Garden within the settlement of Jarasingha

Table 6.18   Plot-wise crop details during the Summer season 2017, Jarasingha

Sl. No.	Summer crops	Plot ID (GIS)	Position of land	Area (in Ha.)	Farmer name
1	Chili	2172_12B	Kitchen garden	0.02	Anirudha swain
2	Mixed crop (cucumber + ridge gourd)	2172_12A		0.01	
3	Bitter guard	2279–2	Outside settlement	0.03	Deba Pradhan
4		4415	Outside settlement	0.03	NA
5	Leafy vegetable	4798A1	Kitchen garden	0.02	ArunaSahoo
6		1862A	Kitchen garden	0.01	Basumathi Sahoo
7		1862B		0.01	
8		2779A	Outside settlement	0.01	Gadadhar Mudali
9		4190		0.03	
10		2172_11A1	Kitchen garden	0.005	Gadadhar Pradhan
11		2172_11A2		0.01	

(continued)

Table 6.18 (continued)

Sl. No.	Summer crops	Plot ID (GIS)	Position of land	Area (in Ha.)	Farmer name
12	Leafy vegetables	2172_11A3	Kitchen garden	0.01	Gadadhar Pradhan
13		2172_11A5		0.003	
14		2172_11A7		0.004	
15		2172_11A8		0.004	
16		4780		0.02	Girish swain
17		4785A		0.01	
18		4785 AD		0.003	
19		4785B		0.01	
20		2034A2		0.002	Lambodhar Parida
21		2034A1_1		0.01	Late Ratnakar Sahoo
22		2034A1_2		0.01	
23		2034A1_3		0.01	
24		2034A1_5		0.01	
25		2034A1_6		0.01	
26		2164	Outside settlement	0.06	Jhnanavi Sahoo
27		1611_A	Kitchen garden	0.02	Madan swain
28		2159_A		0.03	
29		2159_B		0.01	
30		1611	Outside settlement	0.17	NA
31		1667B	Kitchen garden	0.02	
32		1679		0.04	
33		1708A		0.02	
34		1755A		0.01	
35		1977A2		0.01	
36		2016A		0.01	
37		2034A1		0.01	
38		2034A3_1		0.003	
39		2034A3_2		0.003	
40		2090		0.04	
41		3707A		0.02	
42		4211_42		0.15	
43		4417		0.05	

(continued)

Table 6.18 (continued)

Sl. No.	Summer crops	Plot ID (GIS)	Position of land	Area (in Ha.)	Farmer name
44	Leafy vegetables	4699C	Kitchen garden	0.003	NA
45		2281B	Outside settlement	0.02	Narayana Sahoo
46		4793_1	Kitchen garden	0.01	Prafulla swain
47		4778-2B	Outside settlement	0.01	Ramesh Parida
48	Spinach	4785C	Kitchen garden	0.01	Girish swain
49		2034A1		0.005	Late Ratnakar Sahoo
50		1667		0.02	NA
51		1977A3		0.01	
52		2064		0.03	
53		2070		0.01	
54		2200		0.04	
55	Lady's finger	2839	Outside settlement	0.02	JaduBehra
56		4964		0.05	Janardhan Sahoo
57		2278		0.05	Kailash Pradhan
58		2279–3		0.01	Kmar Sahoo
59	Lady's finger	1507–2	Outside settlement	0.07	NA
60		1755	Kitchen garden	0.002	
61		2302	Outside settlement	0.08	
62		2317		0.03	
63		4975	Kitchen garden	0.03	
64		2274–2	Outside settlement	0.05	Narayana Sahoo
65		2281		0.03	
66		2281A		0.04	
67		2279–4		0.02	Pandu Sahoo
68		1755A	Kitchen garden	0.01	Prakash Chandra swain
69		2289	Outside settlement	0.06	Srikanta Pradhan

(continued)

Table 6.18 (continued)

Sl. No.	Summer crops	Plot ID (GIS)	Position of land	Area (in Ha.)	Farmer name
70	Pudina	1670	Kitchen garden	0.03	NA
71	Brinjal	1828		0.02	
72		2172_11A		0.01	Gadadhar Pradhan
73	Beans	2063_1		0.002	NA
74		4821_2		0.03	
75		2322A	Outside settlement	0.02	
76	Pumpkin	2116–1	Outside settlement	0.04	
77		4778–1		0.03	
78		2116-2B		0.01	Ramesh Parida
79		2126	Kitchen garden	0.04	NA
80		2172–9		0.05	
81	*Til*	3096	Outside settlement	0.11	
82	Ridge gourd	4875–6	Kitchen garden	0.03	
83		4970	Outside settlement	0.11	Bijay Ku. Sahoo
84		1866–3	Kitchen garden	0.01	NA
85	Snake gourd	4793_2		0.01	Prafulla swain
86		2172-8A		0.03	Prakash swain
87		4168-1HA	Outside settlement	0.02	Lavanya Mudali
Total				2.25	

Picture 6.16 Geographical coverage of *Kharif* Crops (September 2017) in Jarasingha

Picture 6.17 Crops in *Kharif* (2017) season in Jarasingha

Table 6.19 Jarasingha Village *Kharif* crop's (September 2017) details

Cropping pattern	Category	Crop name	Total No of plots	Area in ha	Area different category (Ha.)	%
Single crop	Cereals	Maize	12	0.51	142.19	74.5
		Paddy	3116	141.61		
		Raagi	3	0.07		
	Pulses	Black gram	4	0.11	10.23	5.4
		Tur	145	10.12		
	Oil seeds	Groundnut	12	0.72	0.72	0.4
	Tubers	Turmeric	4	0.39	0.39	0.2
	Vegetable crops	Beans	3	0.06	13.81	7.2
		Brinjal	5	0.17		
		Carrot	1	0.06		
		Cauliflower	288	7.95		
		Cowpea	3	0.07		
		Cucumber	15	0.53		
		Ginger	9	0.18		
		Lady's finger	41	1.33		
		Leafy vegetable	108	3.24		
		Ridge gourd	5	0.15		
		Saru	1	.0.01		
		Snake gourd	2	0.05		
		Spinach	1	0.01		
Mixed crop	Cereals and vegetable crops	Maize, cowpea, and lady's finger	1	0.05	0.58	0.3
	Oil seeds + pulses	Groundnut+ tur	3	0.15		
	Tubers + vegetable crops	Ginger + cauliflower	1	0.05		
	Vegetable crops	Brinjal + lady's finger	1	0.02		
		Leafy vegetable and cauliflower	1	0.02		
		Mixed vegetable (lady's finger, leafy vegetable, cauliflower)	11	0.29		
	Perennial crops	Banana	2	0.06	13.28	7.0
		Mango plantation	34	11.58		
		Coconut	9	0.47		
		Teak plantation	5	1.17		
	Perennial + understory	Mango plantation + cauliflower	1	0.06	1.62	0.8
		Mango plantation + maize + beans	1	0.36		
		Mango plantation + tur	7	1.20		
	Plantations	Mixed plantation	38	7.92	7.92	4.2
Total			3893	190.74	190.74	100.0

Table 6.20 Overall cropping pattern in the village of Jarasingha

Cropping pattern	Category	Crop name	Rabi Area in Ha.	Rabi Aggregate Area (Ha.)	Summer Area in Ha.	Summer Aggregate Area (Ha.)	Kharif Area in Ha.	Kharif Aggregate Area (Ha.)
Single crop	Cereals	Maize		0.00		0.00	0.51	142.19
		Paddy					141.61	
		Raagi					0.07	
	Pulses	Black gram	16.54	32.84		0.00	0.11	10.23
		Tur					10.12	
		Green gram	16.06					
		Horse gram	0.24					
	Oil seeds	Groundnut	1.19	1.28		0.11	0.72	0.72
		Mustered	0.09		0.11			
	Tubers	Turmeric		5.53	0.00		0.39	0.58
		Ginger			0.00		0.18	
		Garlic	0.81		0.00			
		Onion	3.25		0.00			
		Potato	1.47		0.00			
	Vegetable crops	Beans	0.33	48.04	0.06	2.06	0.06	13.63
		Brinjal	0.14		0.02		0.17	
		Bitter guard	0.07		0.07			
		Carrot			0.00		0.06	
		Cabbage	0.19		0.00			
		Cauliflower	1.08		0.00		7.95	
		Chili	0.02		0.02			
		Cluster beans	0.04					
		Coriander	0.40					
		Cowpea					0.07	
		Cucumber	1.99				0.53	
		Green peas	37.84					
		Lady's finger	0.23		0.52		1.33	
		Leafy vegetable	4.95		0.93		3.24	
		Ridge gourd	0.03		0.15		0.15	
		Saru			0.00		0.01	
		Snake gourd			0.05		0.05	
		Spinach			0.12		0.01	
		Palak	0.04					
		Pumpkin	0.44		0.10			
		Pudina			0.03			
		Radish	0.07					
		Tomato	0.18					
	Fodder crops	Fodder	0.24	0.37		0.00		0.00
		Fodder *chani*	0.13					

(continued)

Table 6.20 (continued)

Cropping pattern	Category	Crop name	Rabi Area in Ha.	Rabi Aggregate Area (Ha.)	Summer Area in Ha.	Summer Aggregate Area (Ha.)	Kharif Area in Ha.	Kharif Aggregate Area (Ha.)
Mixed crop	Cereals and vegetable crops	Maize, cowpea, and lady's finger		0.00	0.00	0.11	0.05	0.59
	Oil seeds + pulses	Groundnut+ tur			0.00		0.15	
	Tubers + vegetable crops	Ginger + cauliflower			0.00		0.05	
	Vegetable crops	Brinjal + lady's finger			0.00		0.02	
		Leafy vegetable and cauliflower			0.00		0.02	
		Leafy vegetable + pumpkin			0.09			
		Cucumber+ ridge gourd			0.01			
		Mixed vegetable (lady's finger, leafy vegetable, cauliflower)			0.00		0.29	
	Perennial crops	Banana	0.06	14.90	0.09	14.93	0.06	13.28
		Mango plantation	13.20		13.20		11.58	
		Coconut	0.47		0.47		0.47	
		Teak plantation	1.17		1.17		1.17	
	Perennial + understory	Mango plantation + cauliflower			0.00		0.06	1.62
		Mango plantation + maize + beans			0.00		0.36	
		Mango plantation + tur			0.00		1.20	
	Plantations	Mixed plantation	7.75	7.75	7.80	7.80	7.92	7.92
Total			110.71	110.71	25.01	25.01	190.76	190.76

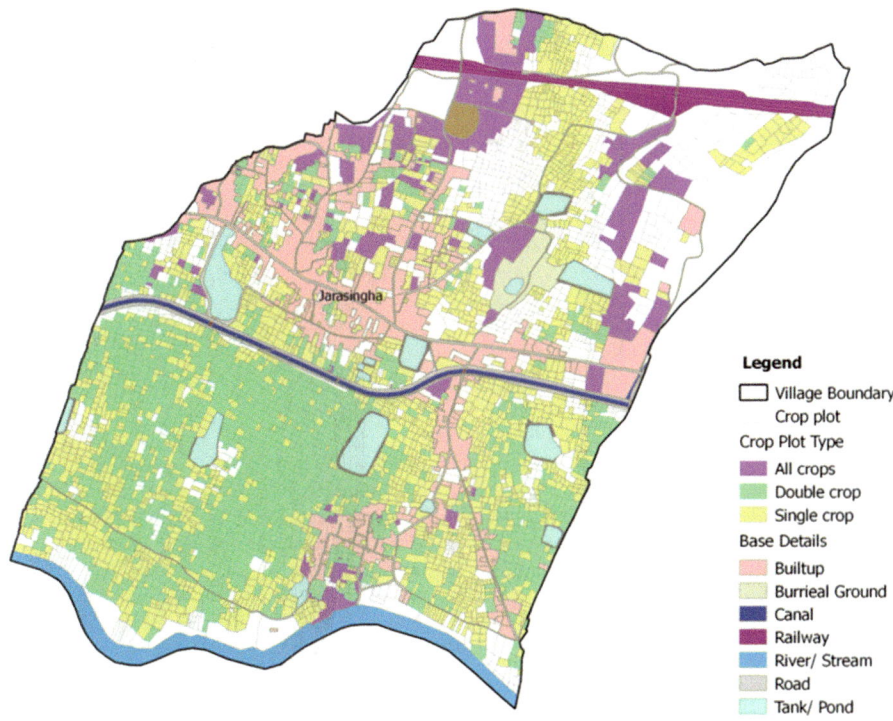

Picture 6.18   Crop plots of all seasons in Jarasingha

Table 6.21 Plots cultivated during the year 2016–2017 and Sept. 2017 in the village of Jarasingha

Type	No of plots	Total area in ha	Crops grown (*Rabi*, summer crops (2016–2017) and with *Kharif* (Sept. 2017)	Reasons
All season plots	148	24.24	Mango plantation, coconut, banana, black gram, leafy vegetable, paddy, brinjal, snake gourd, cauliflower, cabbage, pumpkin, coriander, lady's finger, spinach, mixed vegetable (lady's finger, leafy vegetable, cauliflower), cucumber, garlic, leafy vegetable + pumpkin, tur, ridge gourd, green gram, green peas, beans, groundnut, bitter guard, pudina, mango plantation + cauliflower, mango plantation + maize + beans, mango plantation + tur, onion, ridge gourd, teak plantation	Cultivated by using canal water and groundwater
Two season plots	1770	80.49	Paddy, green gram, black gram, green peas, lady's finger, spinach, beans, cauliflower, leafy vegetable, ridge gourd, groundnut, brinjal, cabbage, cowpea, cucumber, maize, cowpea and lady's finger, snake gourd, chili, (mixed vegetable (lady's finger, leafy vegetable, cauliflower)), cluster beans, coriander, fodder chani, garlic, horse gram, pumpkin, black gram, spinach, ridge gourd, snake gourd, mixed crop (cucumber+ ridge gourd), mustered, onion, palak, potato, pumpkin, bitter guard, snake gourd, ginger, tomato, tur, turmeric	Paddy crop cultivated by canal water in *Kharif*. Black gram, green peas, and green gram are cultivated in *Rabi* season using soil moisture, and vegetable crops are cultivated by using groundwater
Single season plots	2146	93.07	Paddy, black gram, green gram, green peas, lady's finger, leafy vegetable, leafy vegetable + pumpkin, beans, bitter guard, brinjal, brinjal + lady's finger, cabbage, carrot, cauliflower, coriander, cowpea, cucumber, fodder, garlic, ginger, ginger + cauliflower, groundnut, groundnut+ tur, lady's finger, leafy vegetable + cauliflower, maize, mixed vegetable (lady's finger + leafy vegetable + cauliflower), mustered, onion, potato, pumpkin, raagi, radish, ridge gourd, Saru, tomato, tur, turmeric	Cultivated by canal and rainwater. Vegetable crops are cultivated by using groundwater (open well and borewell)

Chapter 7
Reasons for Noncultivation

Abstract Reasons like social, economic, and natural make farmers to keep the land uncultivated. Study highlights not a single but combination of reasons in different regions. In Khatijapura, the scarcity of water and labor are important reasons for not cultivating during Kharif, while the traditional practice also appears to be the reason. In Daliparru, the intention to convert the land to non-agriculture and migration to urban area is the major reason than shortage of water for cultivation. In Jarasingha, shortage of water only appears to be the major reason for leaving the land vacant.

Keywords Social · Economic · Natural causes · Scarcity of water
· Non-agriculture land · Migration

7.1 Why Do Farmers Not Cultivate?

Generally, the farmers do not keep the productive land vacant without cultivation, though temporarily during one of the seasons the land may be kept fallow. Different agroclimatic regions have varying situations, and practices of cultivation differ, and commonly the sources of water determine the pattern and extent of cultivation. In the recent times urbanization coupled with economic constraints and labor issues appears to have left substantial extent of land uncultivated. The reasons vary from region to region.

It was common during *Rabi* and *Summer* that a substantial number of crop plots were not cultivated, particularly in Khatijapura. Similar situation also exists in other two villages. Though farmers were enquired during the earlier inventory, it was felt to introduce a feature in the Android application so that responses, if any, could be recorded on the spot itself. The following reasons for not cultivating the land have been included in the mobile application itself:

1. Water shortage
2. Seed problem
3. Labor problem

4. Financial problem
5. No one to take care
6. Migrated to town/city
7. Not interested (too big/too small area)
8. Non-economical
9. Distance from home
10. Traditional practice
11. Water logging/salinity problem
12. Intent to convert for nonagricultural purpose

7.2   Reasons for Noncultivation

During the field inventory, farmers were asked the questions as indicated above, and the response was recorded. The results of this study in each village are provided below:

7.2.1   Khatijapura

During the *Kharif* (August 2017), 201 plots were uncultivated which comprises of 57% of total cultivable land. The responses from the farmers are compiled in Table 7.1:

- 39% of farmers who have not cultivated their plot during *Kharif* season because of water shortage and labor problem.
- 21% of farmers indicated that not cultivating during *Kharif* season is a traditional practice.
- Both labor problem and traditional practice are the reasons for not cultivating during *Kharif* season for 20% of farmers.
- 7% farmers opined that water shortage, labor problem, and traditional practice are the reasons for not cultivating during *Kharif* season.
- 4% farmers informed that labor problem and farmer's migration to nearby city/ town are the reasons for not cultivating during *Kharif* season.
- 6% of farmers indicated that labor problem is the sole reason for not cultivating their plots during *Kharif* season.
- 3% farmers quoted that water shortage is the sole reason for not cultivating during *Kharif* season.
- Water shortage and traditional practice are the reasons for not cultivating their plots during *Kharif* season for 1% of farmers (Fig. 7.1).

Table 7.1 Reason for not cultivating, Khatijapura

S. No.	Reason for not cultivating	No. of farmers	No. of plots	Area in Ha.	% of farmers	% of plots
Sole reason						
1	Water shortage	3	3	5.83	3	1
2	Seed problem	0	0	0.00	0	0
3	Labor problem	7	8	10.50	6	4
4	Financial problem	0	0	0.00	0	0
5	No one to take care	0	0	0.00	0	0
6	Migrated to city/town	0	0	0.00	0	0
7	Not interested (too big/too small area)	0	0	0.00	0	0
8	Non-economical	0	0	0.00	0	0
9	Distance from home	0	0	0.00	0	0
10	Traditional practice	24	48	75.43	21	24
11	Water logging/salinity problem	0	0	0.00	0	0
12	Intent to convert for nonagricultural purpose	0	0	0.00	0	0
Multiple reasons						
1	Labor problem + traditional practice	23	30	49.65	20	15
2	Labor problem + migrated to town/ city	5	8	18.62	4	4
3	Water shortage + labor problem	45	85	97.79	39	42
4	Water shortage + traditional practice	1	1	0.45	1	0
5	Water shortage + labor problem + traditional practice	8	18	21.97	7	9
Total		116	201	280.24	100	100

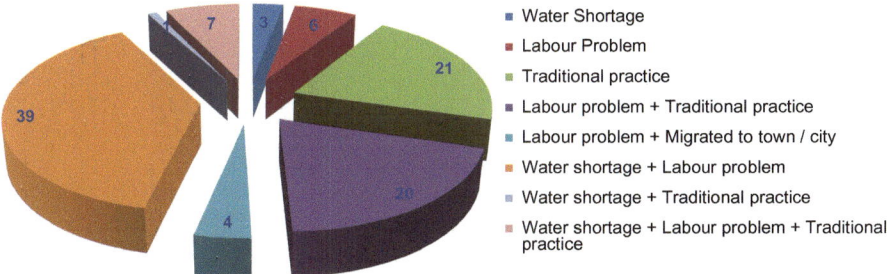

Fig. 7.1 Reasons for not cultivating the plots during *Kharif* in Khatijapura

7.2.2   Daliparru

The entire cultivable lands in Daliparru are under irrigation. Except for a few uplands, the rest of them have water for irrigation. During this season, seven plots were uncultivated which comprises of 0.32% of total cultivable land. Farmers' responses are summarized as follows:

- 71.43% of farmers who have not cultivated their plot during *Kharif* season because of the intent to convert for nonagricultural purpose and migration to city/ town.
- 28.57% of farmers indicated that not cultivating during *Kharif* season is because of water shortage and traditional practice (Fig. 7.2).

7.2.3   Jarasingha

Jarasingha has two district agricultural landscapes – one with canal irrigation facility and other without canal water. A few local tanks in the non-command area supplement water for agriculture for a short duration.

In the command area, there are two types of reasons put forth by the farmers:

- Certain portion of the command area does not have proper field drainage channels, and the land gets waterlogged and therefore left uncultivated. There are 52 plots of this type covering an area of 2.38 Ha.
- Certain uplands in the command area do not get the supply from channels and therefore not cultivated. There are 228 plots covering an area of 11.35 Ha.

In the non-command area, generally the rainfed agricultural lands, due to shortage of water, 717 plots have been recorded with no crop during the *Kharif* 2017 (Picture 7.1; Fig. 7.3).

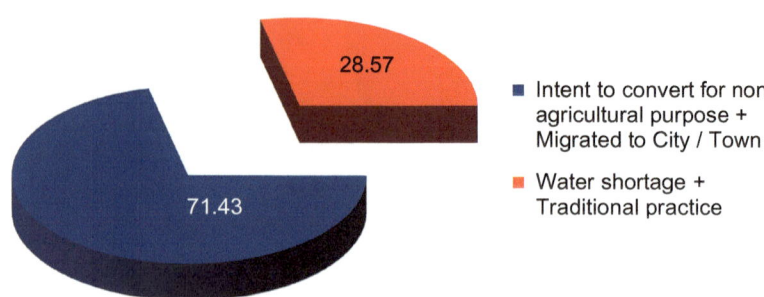

Fig. 7.2 Reasons for not cultivating the plots during *Kharif* in Daliparru

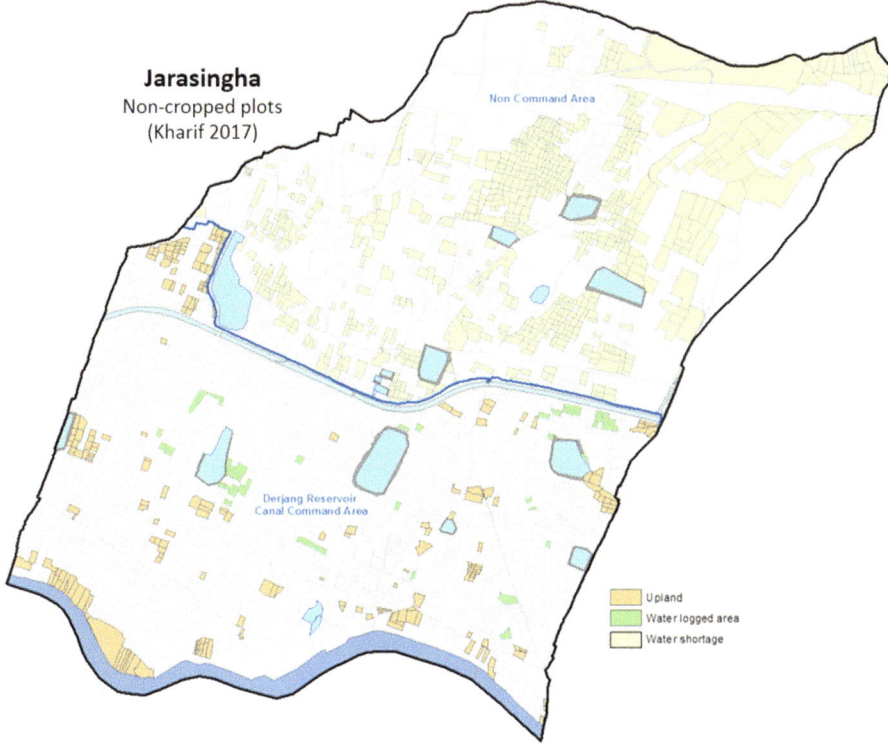

Picture 7.1   Non-cropped plots during *Kharif* in Jarasingha

Fig. 7.3   Uncultivated
plots in Jarasingha

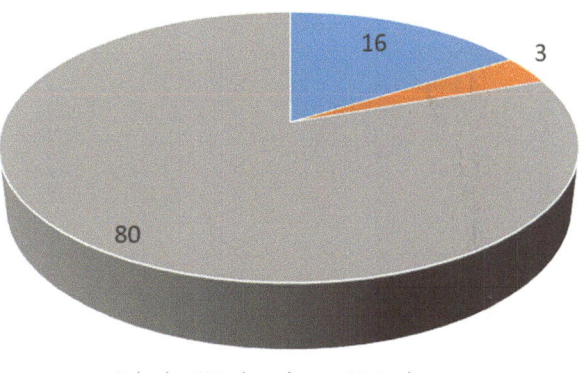

Chapter 8
Land Use

Abstract Land use is the surface utilization of all developed and vacant land on a specific location, at a given time and space, and it defines the human activities which are directly related to land, making use of its resources, or having an impact on them. In that context, the emphasis is on the purpose for which the land is used, and a particular reference is made to "the management of land to meet human needs." Detailed mapping and field authentication using geospatial tools have enabled to accurately define the current land use in three villages. Khatijapura accounts for 92.5% of the net sown area, Daliparru 84.47%, and in Jarasingha less than 50%.

Keywords Surface utilization · Developed land · Vacant land · Current land use surface utilization developed and vacant land geospatial tools

8.1 Importance of Land Use

For any country, toward addressing the development, adequate information on many complex interrelated aspects of its activities is essential to make decisions. Land use is one such aspect, but knowledge about land use and land cover has become increasingly important as the country plans to overcome the problems of haphazard, uncontrolled development, deteriorating environmental quality, loss of prime agricultural lands, destruction of important wetlands, etc. Land use data are needed in the analysis of environmental processes and problems that must be understood if living conditions and standards are to be improved or maintained at current levels.

India has been following two types of land use classification—national land use schema and ninefold classification. Though remote sensing techniques can, to a good extent, help to generate, there is always a chance for misclassification—vegetated vs cropped. Many a times as the old records are reproduced, the changing land use pattern at microlevel is overlooked. Land use at village level is crucial for planning at microlevel. Even the detailed studies using high-resolution satellite images have had limitation in deriving accurate land use at village level. Therefore,

K. V. Raju et al., *Geospatial Technologies for Agriculture*, SpringerBriefs
in Environmental Science, https://doi.org/10.1007/978-3-319-96646-5_8

systematically mapping and authenticating the use of each parcel are important for determining the accurate land use.

One of the most useful outcomes of the intervention has been updating of the land utilization pattern in all the three villages. It was possible because of the following reasons:

- Accurate base map was prepared using high-resolution satellite data that was neatly registered with the village revenue map showing all the revenue holdings including the settlement.
- The base map was converted in to geo-fenced tile and loaded on to GPS-based mobile and every plot was geo-stamped.
- The land use of every plot was recorded with photograph, and the settlement boundary was authenticated in the field itself using the GPS.
- The data was systematically organized in GIS for visualization and analysis.

8.2   Land Use in Three Villages

Land use classification based on the data gathered during the *Rabi* and *Summer* seasons in all the three villages (Table 8.1) has been arrived. Since the third time inventory, essentially of the *Kharif* season, was conducted during August–September 2017, data is not used for land use. The current *Kharif* would constitute the information for 2017–2018 land use. Therefore, the land use derived based on *Rabi* with inferred *Kharif* is retained.

Table 8.1 Land use classification based on the data gathered during the *Rabi* and *Summer* seasons

Code	Land classification	Khatijapura Area in ha	%	Daliparru Area in ha	%	Jarasingha Area in ha	%
1	Forest	0	0	0	0.00	11.65	3.17
2	Land put to nonagricultural uses	30.86	6.43	75.06	12.91	95.73	26.02
3	Barren and uncultivable land	0	0	0.00	0.00	0.00	0.00
4	Permanent pastures and other grazing lands	0	0	0.00	0.00	5.00	1.36
5	Miscellaneous tree crops and other groves, not included in net area sown	1.62	0.29	0.00	0.00	0.00	0.00
6	Cultivable waste	0	0	2.47	0.42	17.84	4.85
7	Fallow land other than current fallow	0	0	1.31	0.23	61.92	16.83
8	Current fallow (*Rabi* fallow included)	3.45	0.72	11.45	1.97	67.01	18.22
9	Net area sown	442.49	92.56	490.94	84.47	108.72	29.55
Total		478.42	100.00	581.23	100.00	367.87	100.00
Water features							
	Stream/river	6.29				9.86	
	Pond/tank	0.18		12.89		12.33	
	Canal + canal ROW			23.68		3.20	
Net area sown							
	Annual	1.58		0		0.00	
	Kharif	145.31		0		0.00	
	Perennial	2.59		0.90		20.38	
	Rabi	293.01		490.04		88.34	
	Total	442.49		490.94		108.72	

Chapter 9
Repetition in Second Year

Abstract Crop inventory experiment was continued during the second year (Rabi season of 2017–2018) to evaluate the feasibility of transferring the data to the server on real time and involving local youths in the process. Spatial reference database of crop plots and holding had hardly any changes and was conveniently used. Internet signal was fine at the village center and data could be directly transmitted to the server, and monitoring the field inventory was achieved. Crop data of Rabi season for 2 years could be conveniently compared. Local youths could comfortably handle the mobile devices and capture the data and synchronize with the server.

Keywords Crop inventory · Internet signal · Village center · Mobile devise · Local youths · Synchronize

9.1 Objectives and Preparation

In order to evaluate the efficiency of the process and to monitor the inventory on real time, the experiment was continued in one of the three villages. Khatijapura village in Karnataka was selected for the inventory during *Rabi* season (2017–2018). Also it was intended to explore possibility of training/involving local youths for the inventory. The villagers were contacted to understand the field situation so as to plan the window for the field session. Accordingly, the field inventory was conducted during early February 2018.

The spatial database of crop plots and the holding details developed during the earlier round of inventory in GIS format was used. The desktop GIS was upgraded to online GIS. The database structure and application architecture were suitably modified to enable direct synchronization of the field data with online server. Also the Android Application loaded on the smartphone was also modified.

Since the local farmers were acquainted with field enumeration team during the earlier rounds, the cooperation and involvement of people were appreciable, and few local youths (Picture 9.1) came forward to capture the data using smartphones.

Picture 9.1 Local youth
capturing the data using
smartphone in Khatijapura

A brief training session was also held for local youths and senior farmers who
showed keen interest in the work. The enumerators moderated with local youths to
capture the data in the field, and the survey was completed within 2 days.

The data captured in the field was locally stored in the smartphone as the internet
connectivity was not strong in all the locations of the village. As the internet connectivity was available in Khatijapura settlement, the data stored in the smartphone
was synchronized with the GIS server. The backend technical team working on the
online GIS application could confirm the data synchronization.

As the application was customized to receive the data from mobiles and directly
integrating the data with standard database, the data was inserted automatically in
the designated format. The records captured on the field on any day could be visualized in the GIS (Picture 9.2). It helps not only to monitor the field enumeration
progress but also visualize the correctness of the data captured. The data could be
neatly visualized as per the cropping season, crop varieties, and other features as
was customized in the application (Picture 9.3).

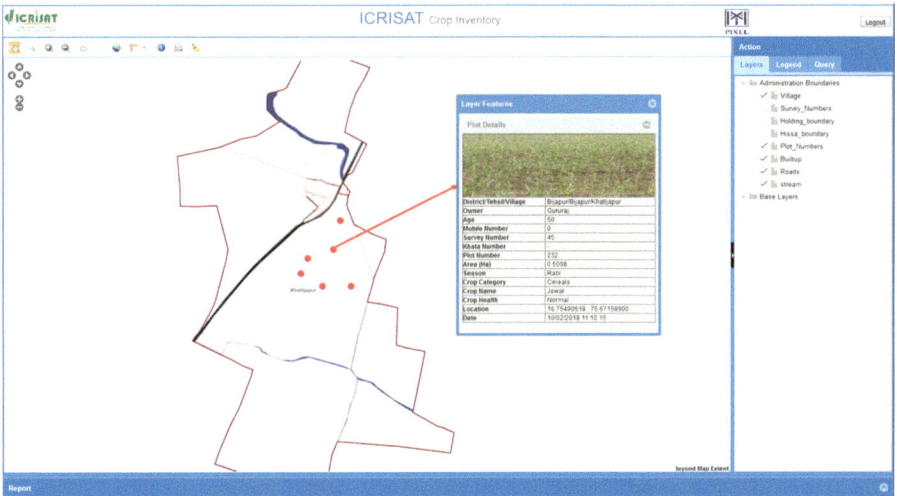

Picture 9.2   Visualization of the locations and details of field data capture in online GIS

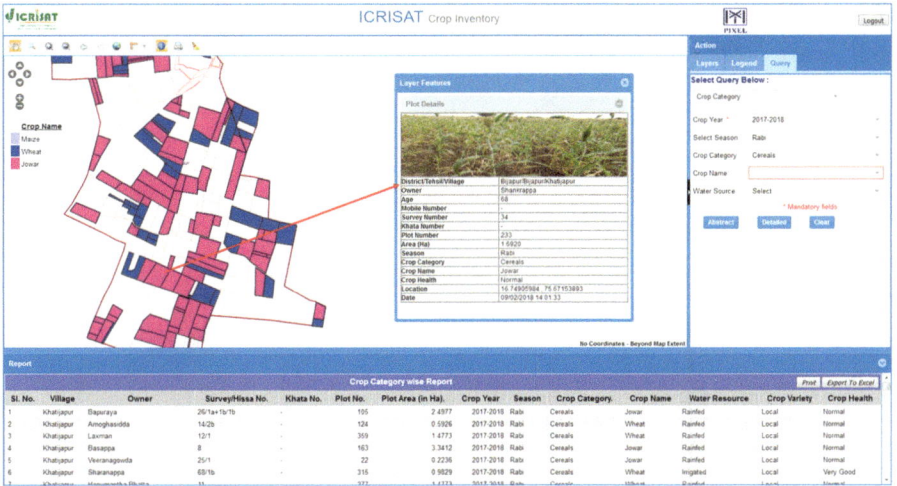

Picture 9.3   Visualization of the crop details in online GIS

9.2   Results of Repeated Inventory

During inventory, 350 plots were inventoried. Out of which 253 plots had standing *Rabi* crops. The entire village had single-crop pattern during the *Rabi*, and crops enumerated in *Rabi* season are shown in Table 9.1 and Pictures 9.4 and 9.5.

The extension of the experiments during *Rabi* season of year 2017–2018 has provided information on the varying crop acreage as well as varieties. This

Table 9.1 Crops inventoried during February 2018 in Khatijapura

Crop category	Crop name	No. of plots	Area in ha
Cereals	*Jowar*	94	128.04
	Maize	1	0.86
	Wheat	55	48.72
Pulses	Bengal gram	87	121.49
Oil seeds	Safflower	3	13.10
Vegetable crops	Brinjal	3	1.71
	Chili	1	0.93
	Cluster bean	1	0.38
	Cucumber	3	0.436
	Ridge guard	1	0.34
Flowers	Chrysanthemum	2	0.16
Fodder crops	Fodder maize	2	1.20
Total		253	317.35

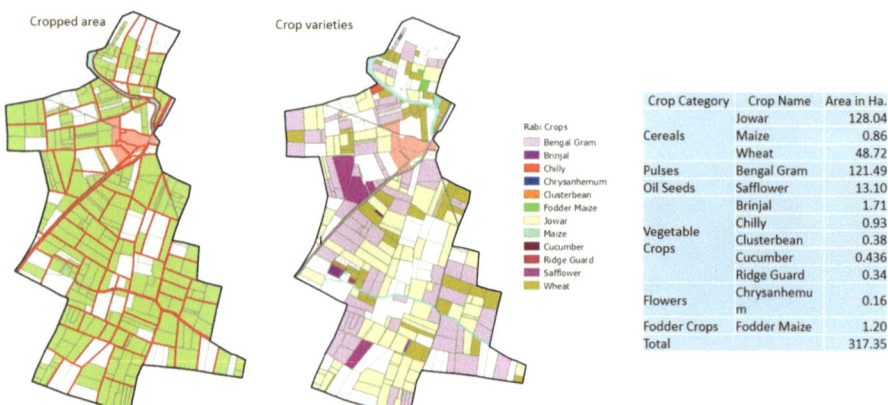

Picture 9.4 Cropping pattern in Khatijapura, *Rabi* 2017–2018 (February 2018)

information is very much useful in understanding the cause leaving land as seasonal fallows or using the same in consecutive seasons.

Further, usage of the mobile application for capturing the data by the local youths was encouraging. It was a good indication that given the opportunity, the local populace is always interested in adopting newer technological interventions.

Variation of Crops during Rabi 2017 and 2018 in Khatijapura

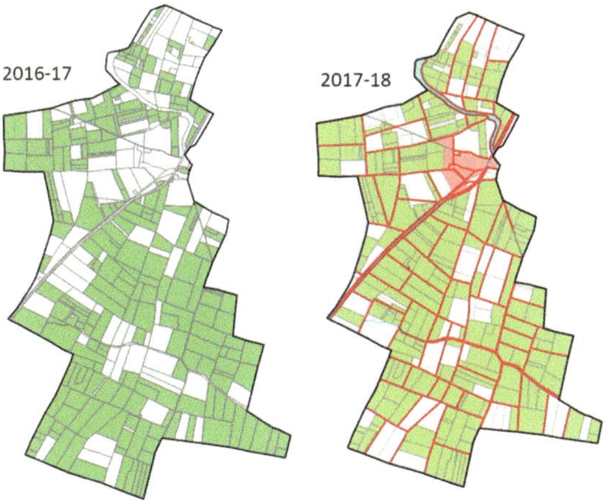

Khatijapura – Rabi Season: 2016-17 and 2017-18 Extent in Ha.		
Crop	2016-17	2017-18
Blackgram	2.99	
Chilly	0.12	0.93
Beans	0.34	0.38
Ridge Guard		0.34
Bengalgram	97.21	121.49
Bengalgram + Jowar	7.99	
Brinjal	1.06	1.71
Chrysanthemum	0.38	0.16
Cucumber	0.26	0.44
Jowar	125.66	128.04
Maize	2.91	0.86
Safflower	10.13	13.10
Wheat	42.15	48.72
Wheat + Jowar	1.29	
Fodder maize	0.25	1.20
Napier grass	0.27	
Total	293.01	317.35

Picture 9.5 Variation of acreage and varieties of crops during *Rabi* of different years in Khatijapura

Chapter 10
Kitchen Garden in Jarasingha

Abstract Jarasingha village has unique feature of Kitchen Garden within settlement attached to dwelling units that supplements livelihood option for the people. Women in the households cultivate different varieties of vegetables using groundwater. Water from borewells is shared between households. Plots vary in size from 10×20 m to 15×30 m, and cumulative extent of such garden is 6.03 ha, and there is a clear shift in crops from leafy vegetables during summer to cauliflower in Kharif mostly due to local market demand. During summer, people practice organic cultivation; while in Kharif, chemical fertilizers and pesticides are used.

Keywords Kitchen Garden · Livelihood option · Vegetables · Organic cultivation · Pesticides

10.1 The Practice

Jarasingha village offered an important feature of Kitchen Garden within settlement. During the *Rabi* season, small garden attached to households in the periphery, inter-unit spaces was observed. Also few tiny gardens within household premises deep in the settlement were noticed. These features were included within the area occupied by the settlement. Informal inquiry with few people indicated that growing vegetables within household premises are a regular practice, and many are engaging in selling the produces also. Therefore, an attempt was made during the summer season to capture more information about this practice.

Detailed study of the satellite image (Google Earth, Picture 10.1) indicates that the settlement is concentrated along the main road from Angul and has extended parts along the roads leading away from the villages. There must have been sufficient agricultural interspaces (which are now slowly covered with houses with such Kitchen Garden) between these roads (mostly on the northern part of the village) that did not have canal irrigation source. People have dug few open wells and started using the water. Slowly, borewells got introduced, and currently, most of them use

Picture 10.1 Image of Jarasingha showing Kitchen Garden (interviewed)

borewell water. Many of the open wells have gone dry. Interestingly, the borewell water is shared between two and three households for such gardening. Generally, each household has 10 × 20 m or slightly larger plots and portion of which is built, and the rest is used for growing different types of vegetables that have influenced on the livelihood of the farmers. The produce from these gardens is sold locally and sent to Angul city also.

10.2   Detailed Inventory

During summer inventory, feasibility of mapping Kitchen Garden was explored. As many as 61 plots were captured, and cumulative area of such garden crops within the settlement is 1.08 ha. An effort was made to interview about 28 farmers (households with Kitchen Garden) about the practice. This initial survey has provided insights in to the practice of Kitchen Garden (Picture 10.2) as a supplementing livelihood option for the people. As the team had not gone prepared with detailed map, all such gardens could not be mapped and interviewed.

During the *Kharif* (September 2017), detailed map of the village settlement was prepared, and a structured interview schedule was made. Using this reference, complete inventory of the Kitchen Garden was made (Pictures 10.3 and 10.4).

The interview focused on the following aspects:

- Basic information.
- Source of water.

Picture 10.2   Crops in the Kitchen Gardens of Jarasingha (summer–May 2017)

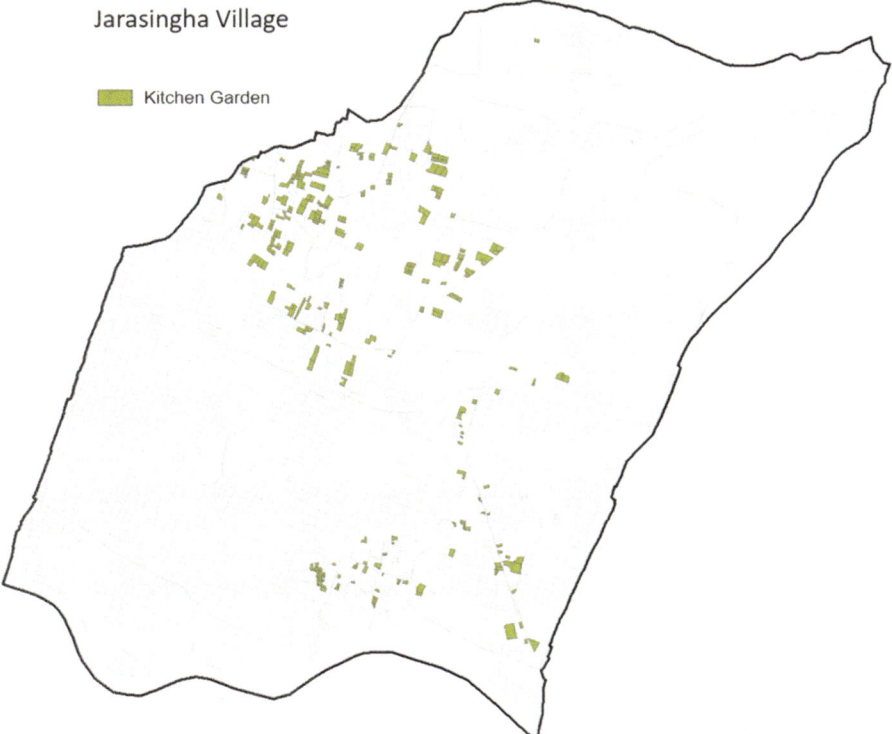

Picture 10.3   Distribution of Kitchen Garden in Jarasingha (Sept 2017)

Picture 10.4 Kitchen Garden during September 2017 in Jarasingha

- Crop details.
- Production details.
- Sales details.
- Seed details.
- Details related to fertilizers, pesticides, and organic manure.
- Labor.
- Expenditure and profit.

10.3   Kitchen Garden: Supplementing Livelihood

Details of the inventory of Kitchen Garden are provided in Table 10.1. In all, 291 plots could be mapped which comes to cumulative area of 6.03 ha. About 17 plots which had crops during summer have been found to be vacant during *Kharif*.

Totally 203 households have been found to be having Kitchen Garden. During the interview, name of the persons/owners of 135 households could be collected. For the rest (71), the information was gathered from the neighbors Table 10.2.

Analysis of the efforts and outputs from Kitchen Garden as shown in the above table indicates the following:

- There is shift in crops from leafy vegetables during summer to cauliflower in *Kharif*. This could be due to local market situations/demand.
- Minimum (holding) extent of plots has been found to be 0.2 Gunta and maximum as 16 Gunta with an average of 3 Gunta.

Table 10.1 Kitchen Garden: Jarasingha (September 2017)

Description	Category	Crop name	No. of plots	Area in ha	Average income per crop in Rs	
					Minimum	Maximum
Kitchen garden	Cereals	Maize	2	0.02	240	250
	Pulses	Tur	7	0.10	340	500
	Tubers	Turmeric	2	0.10	7200	7760
	Tubers + vegetable crops	Ginger + cauliflower	1	0.05	2290	
	Vegetable crops	Beans	3	0.06	115	650
		Brinjal	3	0.08	1150	13,200
		Brinjal + ladies' finger	1	0.02	50	
		Carrot	1	0.06	22,350	
		Cauliflower	151	3.07	450	45,400
		Cucumber	1	0.02	300	
		Ginger	8	0.12	3350	21,500
		Ladies' finger	17	0.36	40	2130
		Leafy vegetable	65	1.44	20	2500
		Leafy vegetable and cauliflower	1	0.02	200	
		Mixed vegetable (ladies' finger, leafy vegetable, cauliflower)	6	0.12	900	1800
		Ridge gourd	3	0.05	70	450
		Saru (Kavale soppu)	1	0.01	270	
		Spinach	1	0.01	480	
Total plots with crops during *Kharif*			*274*	*5.70*		
Plot without crops during *Kharif*			*17*	*0.33*		
Total area of kitchen garden			*291*	*6.03*		
Households with owners	135		*214*	*4.27*		
Households for which information gathered from neighbors	71		*77*	*1.76*		
Source of water		Number of borewells	*29*			
		Number of open wells	*69*			

- Besides these points, during the summer season, people do hardly use any pesticides and fertilizers, while almost all use fertilizers and pesticides during *Kharif*. During summer, a maximum number of people were using organic manure.
- There are about 98 sources of water (29 borewells and 69 open wells) during summer; a maximum number of open wells were found to be dry, while all of them have water during the *Kharif*, and all the borewells are functioning.

Table 10.2 Crop-wise net income per Gunta (Rs.) for one cropping season (Jarasingha)

Kharif (September 2017)			Summer (May 2017)		
Crops	No of households	Income/ Gunta	Crops	No of households	Income/ Gunta
Beans	3	300	Beans	1	3100
Brinjal	3	2100	Brinjal	3	1500
Brinjal and ladies' finger	1	2100	Brinjal and ladies' finger	1	1000
Carrot	1	4000			
Cauliflower	133	3400	Cauliflower	2	
Cucumber	1	310	Kakudi mixed	1	300
Ginger	8	6700			
Ginger and cauliflower	1	5000			
Ladies' finger	17	370	Ladies finger	8	440
Leafy vegetable	62	419	Leafy vegetable	71	1000
Leafy vegetable and cauliflower	1	1900	Leafy vegetable and cauliflower	1	
Leafy vegetable and pumpkin		419	Leafy vegetable and pumpkin	2	
Mixed vegetable	5	1700	Mixed vegetable	6	
Mixed vegetable and cauliflower	1	2500			
Ridge gourd	3	270	Ridge gourd	1	
Snake gourd			Snake gourd	2	760
Turmeric	2	3700			
Note					
Min holding	**Max holding**	**Average**			
0.2	16	3			

Chapter 11
Comparison of Area Statistics: Existing Practice vs Geospatial Technology

Abstract Crop plots derived from satellite images correspond to shape and extent in the field and correlate with land records. Crop statistics generated using geospatial tools are authentic and accurate and traceable but do not correlate with the existing data compiled from traditional methods which have problems in terms of type, extent, and acreage. The land use details also appear to have not been updated. The methodology adopted has been able to provide reliable information on the basic production unit, reasons for not cultivation, farming infrastructure, farming practice, availability of agriculture inputs, and geographic variation of crops. Keeping in view the goal of doubling the income and improved agricultural productivity and climate resilience, geospatial tools appear to be useful for generating the reliable and real-time data. The technology can also be conveniently used for consolidating the spatial information of land records, particularly the holding information.

Keywords Crop plots · Satellite image · Land records · Traceable · Acreage · Farming infrastructure · Climate resilience · Geospatial tools · Real time data

The research experiment in three different agroclimatic regions has been able to generate reliable results on crops, extent of each crop, land use, and cropping practices. Since three states have slightly varying practices of generation of crop statistics, an attempt has been made to compare the results with data available with the respective government departments. The results of comparative analysis are provided in the subsequent sections.

© The Author(s), under exclusive licence to Springer Nature Switzerland AG 2019
K. V. Raju et al., *Geospatial Technologies for Agriculture*, SpringerBriefs
in Environmental Science, https://doi.org/10.1007/978-3-319-96646-5_11

11.1   Correlation of Crop Plots Traced from Images and Field Situations and Revenue Records

An exercise was made to determine the area accuracy of the plots extracted from the satellite image and later registered with village map and statistics generated in GIS. Sample from Daliparru village (where individual sketches were available for reference) is presented below:

The area extracts available from revenue records when compared (Pictures 11.1, 11.2, and 11.3) with crop plots traced from images are found to be matching in the field. The area compiled from GIS and that from the records are presented in the following table (Table 11.1).

The table indicates that the crop plots extracted from high-resolution satellite images can be conveniently used from crop inventory at plot level. The area of each plot or aggregate of plots forming a holding (land of a farmer for which he owns a *khata*) matches (with accuracy of more than 95%). Minor mismatch in the area as per GIS map (currently traversed in the field)—which has been corrected in the GIS environment using UTM projection could also be due to area occupied by field bunds.

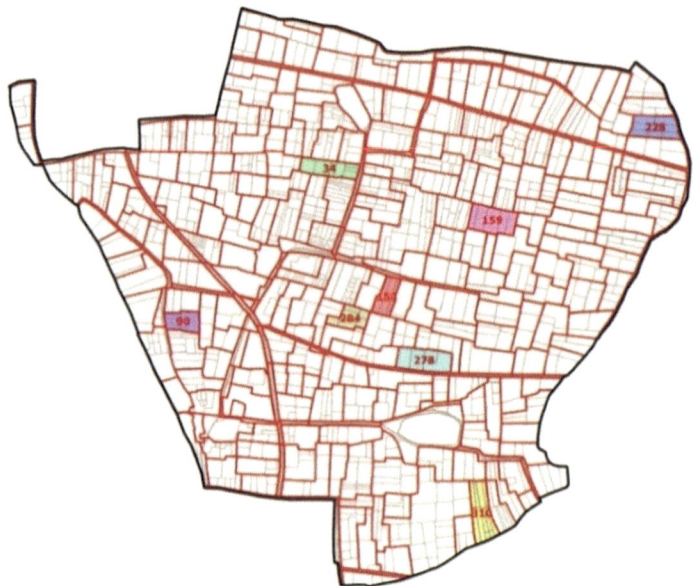

Picture 11.1 Map of Daliparru Village showing plots extracted from image as well as revenue records for comparison

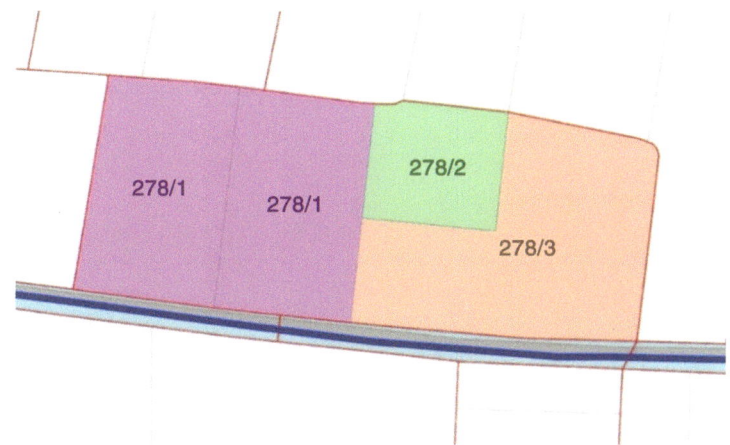

Picture 11.2 Sy. No 278 of Daliparru Village compiled in GIS

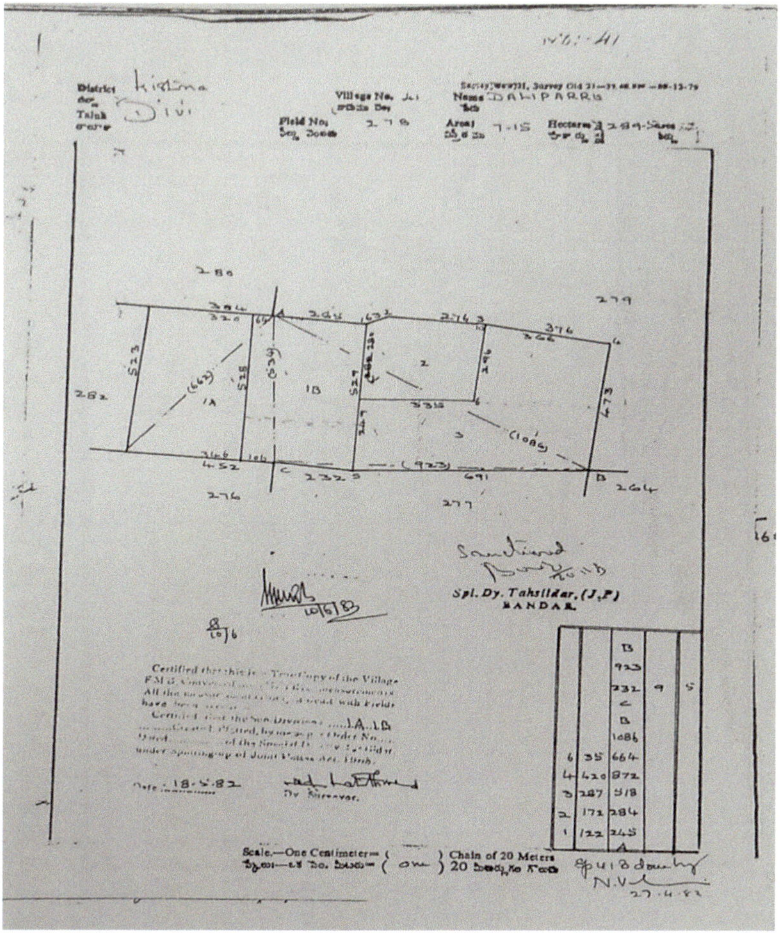

Picture 11.3 Field measurement book (sketch of Sy. No. 278) as per records

Table 11.1 Sample results of the comparison of area obtained from image, field, and revenue records

Sl. No.	Plot Id	Sy. no.	Sy. no Hissa	Khatha No.	Area in acre (extracted from image and compiled in GIS)	Area in acre (as per RTC)	Difference	% of difference
1	1384	150	150-1	258	1.46	1.55	−0.09	−6.00
	801	150	150-2	722	1.69	1.56	0.13	8.13
	1380	150	150-2B	722	1.07	1.00	0.07	6.89
					4.21	4.11	0.10	2.50
2	1422	278	278/1	460	1.80	3.50	0.09	2.57
	1440	278	278/1	620	1.80			
	1607	278	278/2	598	0.94	0.98	−0.04	−3.68
	1608	278	278/3	187	2.77	2.67	0.10	3.81
					7.31	7.25	0.06	0.82
3	1396	284	284/1A	149,155	0.56	0.65	−0.09	−14.50
	1398	284	284/1B	41	0.56	0.56	0.00	0.76
	1395	284	284/1C	224,415	0.96	0.95	0.01	1.53
	1450	284	284/2A	223,222	0.97	1.50	0.12	8.00
	1451	284	284/2A	723	0.65			
	800	284	284/2B	415	0.59	0.54	0.05	8.38
					4.29	4.32	−0.03	−0.69
4	1259	34	34/1	563	1.36	1.43	−0.07	−4.61
	675	34	34/2	367	1.50	1.48	0.02	1.63
	1257	34	34/3	57	0.86	1.45	−0.09	−6.21
	1258	34	34/3	726	0.50			
	1527	34	34/4	57	0.38	0.36	0.02	4.52
	1030	34	34/5	369	0.32	0.34	−0.02	−4.84
	1479	34	34/6	317,336	0.72	0.72	0.00	−0.57
					5.65	5.78	−0.13	−2.32
5	192	90	90-1	80	1.67	1.58	0.09	5.81
	123	90	90/2	362	2.33	2.32	0.01	0.63
					4.01	3.90	0.11	2.73
6	968	228	228-1	51	1.09	2.29	−0.02	−0.87
	970	228	228-1	182	1.18			
	961	228	228-2	707	0.92	1.84	0.10	5.43
	962	228	228-2	447	1.02			
	977	228	228-3	354	0.78	2.28	−0.05	−2.19
	979	228	228-3	353	0.63			
	981	228	228-3	352	0.81			
					6.44	6.41	0.03	0.43

(continued)

Table 11.1   (continued)

Sl. No.	Plot Id	Sy. no.	Sy. no Hissa	Khatha No.	Area in acre (extracted from image and compiled in GIS)	Area in acre (as per RTC)	Difference	% of difference
7	256	310	310-1	444,446	0.51	3.53	0.09	2.55
	355	310	310-1	444,446	0.78			
	238	310	310-1	100,445	1.21			
	481	310	310-1	696	1.12			
	35	310	310-2	446	0.99	1.58	0.01	0.63
	259	310	310-2	510	0.60			
	71	310	310-3	100299	0.49	2.09	0.13	6.22
	72	310	310-3	697	0.56			
	236	310	310-3	697	0.24			
	489	310	310-3	508	0.72			
	490	310	310-3	100,299	0.22			
					7.42	7.20	0.22	3.12
8	1193	159	159-1	123	1.08	1.02	0.06	5.72
	760	159	159-2	6026	1.50	1.53	−0.03	−2.11
	1237	159	159-3	377	2.17	2.22	−0.05	−2.04
	772	159	159-4	676	1.22	1.18	0.04	3.34
	1233	159	159-5	590	1.25	1.2	0.05	4.36
					7.22	7.15	0.07	1.01
							Average	1.08

11.2   Comparison of Crop Statistics

The statistical data pertaining to *Kharif* 2017 was not readily available so that the data gathered by the team could be compared for the respective season. Since the main objective of the experiment was to examine the use and reliability of geospatial technologies, crop—area statistics generated by the existing system—was collected and analyzed. Relevant data were obtained from concerned authorities. The comparative data is presented in the following sections.

11.2.1   Khatijapura

Plot-wise crop details as enumerated using GPS are not available with concerned offices. Holding wise also is not updated. However, some records showing season-wise crop extent were available, and total area sown was made available.

Table 11.2 Comparison of statistics—area sown (Khatijapura)

Year	Kharif (sown area in ha)		Rabi (sown area in ha)	
	Department	GPS	Department	GPS
2015–2016	76	Not done	73	Not done
2016–2017	74	145.31 (inferred in the field)	75	293.01

Table 11.3 Comparison of statistics—crop types (Khatijapura)

Kharif (sown area in ha) 2016–2017		Rabi (sown area in ha) 2016–2017	
Department	GPS	Department	GPS
Bajara	–	Cotton	Sunflower
Tur (red gram)	Tur	Bengal gram	Bengal gram
Green gram	Black gram	Rabi Jowar	Rabi Jowar
Groundnut	– .	Sugarcane (ratoon)	Safflower
Safflower	Safflower mixed with Tur	Sugarcane	Sugarcane
			Pomegranate
			Wheat
			Maize
			Black gram
			Brinjal mixed with onion

11.2.1.1   Area Sown and Crop Types

The Tables 11.2 and 11.3 indicates that the area sown details are almost the same as in the previous year. The geospatial method has been able to generate the statistics from individual plot level and has provided the results that are almost double than the figures provided by the department. Further, the crop varieties also are not properly recorded. Groundnut was recorded during the *Kharif*, whereas it was found during summer inventory and in a small parcel.

11.2.1.2   Land Use

The land use information appears to be not updated in the records maintained by the department. The information obtained from the department (Fig. 11.1) shows area under nonagricultural use as only 5 ha, whereas the plot level inventory revealed it to be 30.86 ha. Substantial extent of cultivable lands has been converted to plots for residential purpose in the surroundings of the village that have remained unoccupied and rendered as scrubland. These might not have been updated in the records.

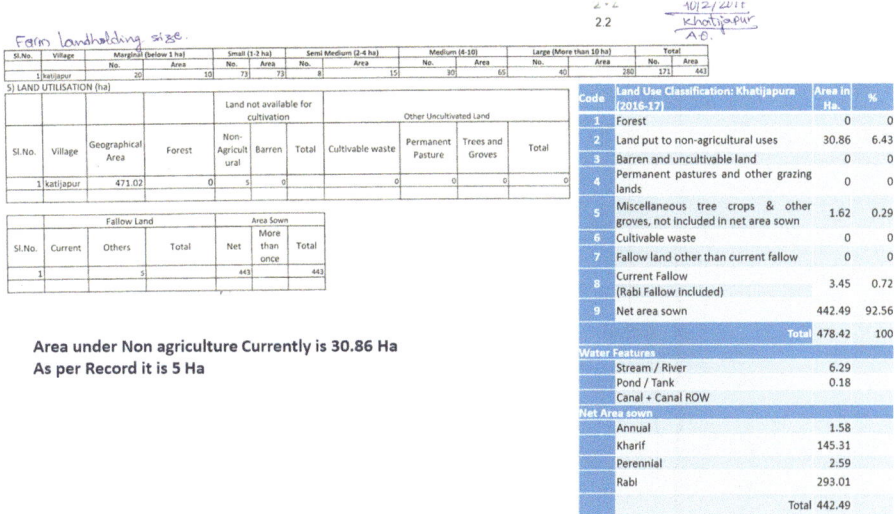

Fig. 11.1 Khatijapura: Mismatch of Land use details – Official records v/s. inventory details

Table 11.4 Comparison of statistics—area, Daliparru

Year	Kharif (sown area in acres, ha)		Rabi (sown area in acres, ha)	
	Department	GPS	Department	GPS
2015–2016	1236 (500.40 ha)	Not done	1234 (499.59 ha)	Not done
2016–2017	1223 (495.14 ha)	Not done	Not provided	501.64 ha

11.2.2 Daliparru

Plot-wise crop details as enumerated using GPS are not available with concerned offices. Holding wise also is not updated. However, aggregate data of the entire village is available for *Kharif*. The main issue here was that data compilation happens late, and information would be available after harvest.

11.2.2.1 Area Sown (Table 11.4)

11.2.2.2 Crop Types

Rabi season (2015–2016) was available, and area of different crops is tabulated against the 2016–2017 *Rabi* inventory as follows (Table 11.5):

Though the variation is not much, it leads to a decision that based on old pattern, the data might have been compiled. Otherwise, minor crop details would also have been available.

Table 11.5 Comparison of statistics—crop types, Daliparru

Crop	Existing system (area in ha.)	RS-GIS-GPS method (area in ha.)
Black gram	496.55	485.71
Green gram	2.83	0.00
Fodder *Jowar*	–	0.15
Gini grass	–	0.39
Horse gram	–	0.15
Others (Saijulu)	–	0.17
Sunnhemp	–	3.36
Turmeric	–	0.11
Total	499.38	490.04

Table 11.6 Departmental data on crops, Jarasingha

Autumn season (July–October)		Winter season (November–February)		Summer season (March–June)	
Paddy	0.50	Paddy irrigated	310.40	–	–
Maize	1.17	Black gram	21.32	–	–
Green gram	3.89	Green gram	132.88	–	–
Black gram	7.43	Other pulses	6.81	–	–
Other crops	1.03	Potato	4.17	–	–
–	–	Other vegetables	27.06	–	–
Total	14.02	Total	502.64	–	–

11.2.3 Jarasingha

The data available from the concerned department is shown below (Table 11.6):

The data is compared with the results obtained from GPS-based inventory which are as follows (Table 11.7):

The winter season corresponding to *Rabi* shows certain difference in the data. The February 2017 GPS inventory does not reveal any paddy crops. Even if paddy is considered to be in the earlier season, acreage of other crops does not correlate. The acreage of other vegetables also has substantial variation (GPS, 47.74 ha.; and department, 10.95 ha.). The general statistics of Jarasingha does not mention the area under Kitchen Garden. As such practices do supplement livelihood options, it may be worthwhile to include survey of such features in the future.

Table 11.7 Comparison of departmental data on crops, Jarasingha

Crop	Area as per existing system (ha.)	Area as per RS-GIS-GPS method (ha.)
Beans	10.95	0.33
Bitter gourd	(other vegetables)	0.07
Brinjal		0.14
Cabbage		0.19
Cauliflower		1.08
Chilly		0.02
Cluster beans		0.03
Cucumber		1.97
Green peas		37.86
Lady's finger		0.23
Palak		0.04
Pumpkin		0.45
Radish		0.07
Ridge gourd		0.03
Leafy vegetable (*Soppu*)		5.05
Tomato		0.18
Black gram	2.76	16.54
Coriander	–	0.4
Fodder	–	0.26
Fodder *chani*	–	0.13
Garlic	–	0.79
Green gram	53.77	16.06
Groundnut	–	1.18
Horse gram	–	0.24
Mustard	–	0.09
Onion	–	3.37
Potato	1.69	1.48
Sesame	–	0.07
Paddy	125.61	0.00
Biri	8.63	0.00
Total	203.41	88.35

11.3 Information Gaps and Advantage of Geospatial Technology Intervention

11.3.1 Reliable Information

Crop statistics has been a debatable issue as the source of information itself and the processes (including the shortcomings) have considerable lacuna. Understanding the constraints of data collection in a fragmented holding, mixed cropping, and seasonal variations, the geospatial technology has been brought in as intervention. The experiment has produced reliable information at plot/holding level in a most transparent, retraceable manner, and the process/intervention can be repeated to produce sets of information for different time domain.

Current practices do not generally generate information on agriculture practices: crop varieties, crop succession, cropping pattern techniques, and the infrastructures. As mentioned earlier, the statistics may be fine at national level or state level. Since the overall goal is toward doubling the income and improved agricultural productivity and climate resilience, the reliable and real-time data becomes important. Unless information at the basic unit of production, i.e., holding of a farmer, the constraints like climatic, economical, or otherwise are not understood at all.

11.3.1.1   Information at Basic Production Unit

Authenticity of the geospatial technology intervention is able to generate most authentic data on crop varieties, crop rotation (succession) at plot level, and the cropping pattern which is difficult to generate either by existing practices or by remote sensing. The varieties of crops grown in an area are accurately registered by the intervention. Practice of changing the crop in different seasons either due to general practice (as is observed in all the three locations) or by choice by the farmer is neatly captured by the current intervention. This is a useful information that provides critical insights in to the scenario at the farm level.

11.3.1.2   Reasons for Not Cultivation

Information about fallow land might be generated by combination of remote sensing images (by selecting multi-season and data at closer intervals provided the data is not masked due to cloud), but during the growing season itself, the information on extent and reasons are understood by the geospatial technology intervention. More than anything else, famer gets a chance to express the local conditions at his farm itself with situations on the ground.

11.3.1.3   Farming Infrastructure

Farmers at the village level due to fragmented and small holding may not be able to bear the cost of supporting infrastructure. The information on this is crucial to understand the requirements at microlevel. Still, many farmers cannot afford to use machineries for land preparation and sowing and/or planting. The method adopted in this study has been able to generate useful information.

11.3.1.4 Farming Practices

Due to various reasons, farmers do not cultivate the land and have issues related to finances. Information at the village level and holding level related to leasing, contract farming, and crop sharing are important which are collected at the farm itself. The method is better than general practices of sampling.

11.3.1.5 Use and Availability of Seeds and Other Inputs

Since every farmer was interviewed at his/her land (holding), useful information related to type of seed he is able to get, the use of fertilizers/pesticides, and support resources has been collected. This would have otherwise not been possible by any other sample surveys.

11.3.1.6 Geographical Aspects of Crop Data

Since the entire information is organized around the GIS, it provides best means of analyzing spatial variations—crops in any given season and changing scenario over a period. Distribution of seasonal fallow lands, permanent fallows, density and distribution of groundwater abstraction structures, etc. are available for visualization.

11.3.1.7 Kitchen Garden

The Kitchen Garden in villages play an important role in the economy of farming household. Generally, this information is overlooked or does not find proper information. The geospatial tools can help best to gather such information.

11.3.2 Update of Land Use

The most useful information is about land use. Though remote sensing techniques can, to a good extent, help to generate, there is always a chance for misclassification—vegetated vs cropped. Many a times the old records are reproduced, and the changing land use pattern at microlevel is overlooked. Land use at village level is crucial for planning at microlevel. Even the detailed studies using high-resolution satellite images have had limitation in deriving accurate land use at village level. One of the most useful outcomes of the intervention has been the update of the land utilization pattern in all the three villages. It was possible because of the following reasons:

- Shape and size of the land parcel were derived from the latest satellite image.
- The integrated parcels were spatially registered with village revenue survey maps.

- Each plot/sub-parcel in a survey number has been correlated with *Khata*.
- Plot level field enumeration was done using GPS.
- The data has been systematically organized around a GIS.

The land use of all the three villages has been derived based on the plot level enumeration. This forms the baseline information as on the year 2016–2017. If this activity is continued, changes over a period of time can be updated.

11.3.3   Consolidated Cadastral Map Providing the Details of Individual Holdings

- It is observed in all the three cases that there is no consolidated village spatial map showing the different holdings boundaries (sub-parcel boundaries). Generally, the revenue records are maintained for each holding along with individual sketch. If an integrated map is generated, it provides an opportunity to correlate the crop details.
- Further, once holding map is available, it becomes easy to visualize the cropping pattern in a village, and it also provides insights into distribution pattern of holdings of different sizes and crops grown during different seasons.
- As the current exercise went into the detailed inventory at crop plot level and collection of information related to practices and the use of fertilizers/pesticides and sources of irrigation, constraints and opportunities for each farmer could be documented. This would go a long way in identifying the type and extent of support for agriculture.
- Holding-wise sketches were not readily available for Khatijapura, whereas very neat maps of individual holdings were available for Daliparru. Jarasingha village appears to be having old and obsolete spatial map of village.

11.3.3.1   Khatijapura

In case of Khatijapura, few farmers were not available for consultation during the field inventory. Overall correlation of the *Khata* and the field plots was possible. The intervention has been able to produce a reliable map of holdings for Khatijapura (Picture 11.4).

11.3.3.2   Daliparru

Revenue records downloaded from the official websites in respect of Daliparru were matching, and the plots could be correlated well with individual *Khata*. The GPS-based inventory and the inputs have been very much useful for identifying the disposition of the *Khata* and crop plots of same *Khata* located at different places in the

Khatijapura

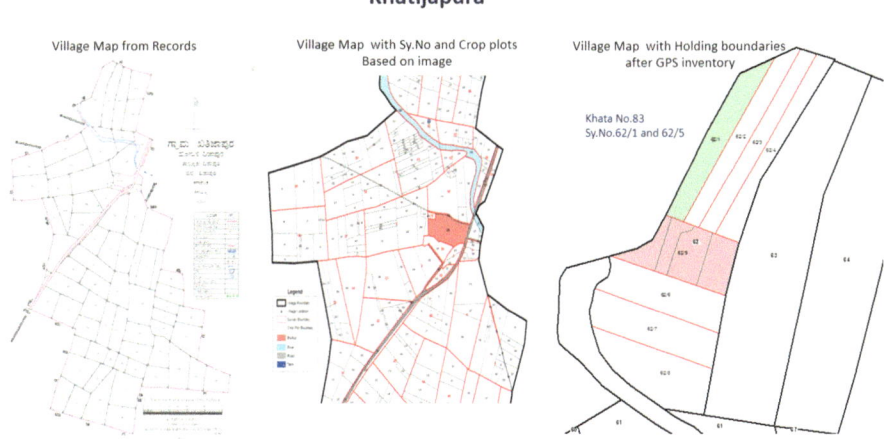

Picture 11.4   GIS-GPS intervention resulting in *Khata*/holding maps of Khatijapura

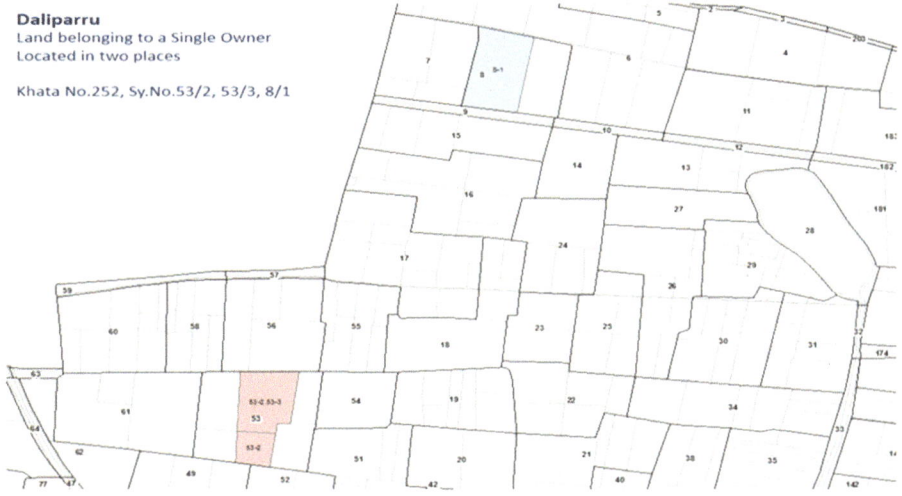

Picture 11.5   GPS inventory providing insights on location of plots of same *Khata* in different places in Daliparru

same village. If the GPS intervention is extended as regular practice, it becomes a most reliable method to derive the crop statistics during the growing season itself, and further, the same methodology can also be used to identify the crop situation (pest attacks/any other problems associated with growth and expected production) (Pictures 11.5 and 11.6).

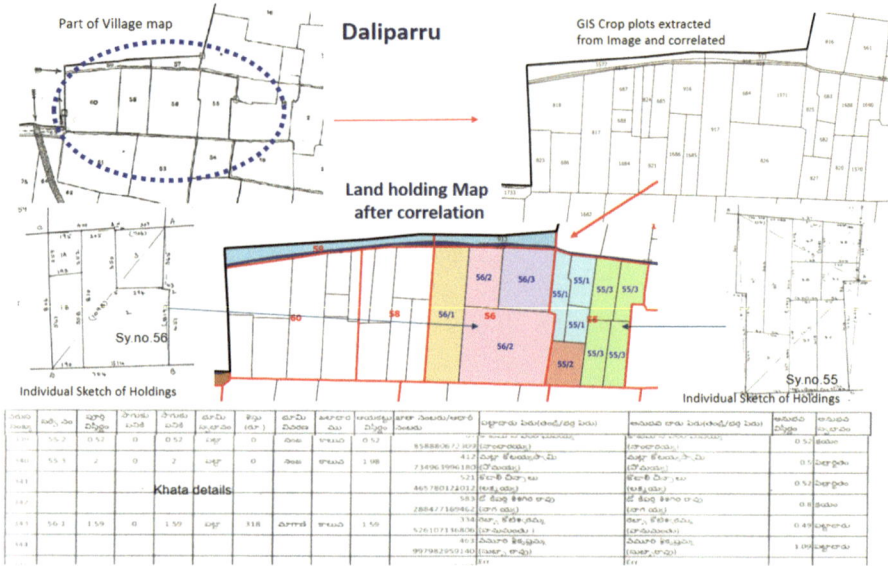

Picture 11.6 Individual holdings in Daliparru correlated with discrete sketches based on GPS enumeration in the field

11.3.3.3 Jarasingha

Map used by the department for crop enumeration does not correlate with the *Khata* extracts (downloaded) from the official website. The plots in the field and that derived from the downloaded village map correlate, and enumeration using GPS was useful. Since farmers did not cooperate during the field inventory, correlation of holdings with farmer name was difficult.

If the maps available in the official records and correlated with field situations are used and enumeration is done, the statistics could be more useful (Pictures 11.7, 11.8, 11.9, and 11.10, 11.11).

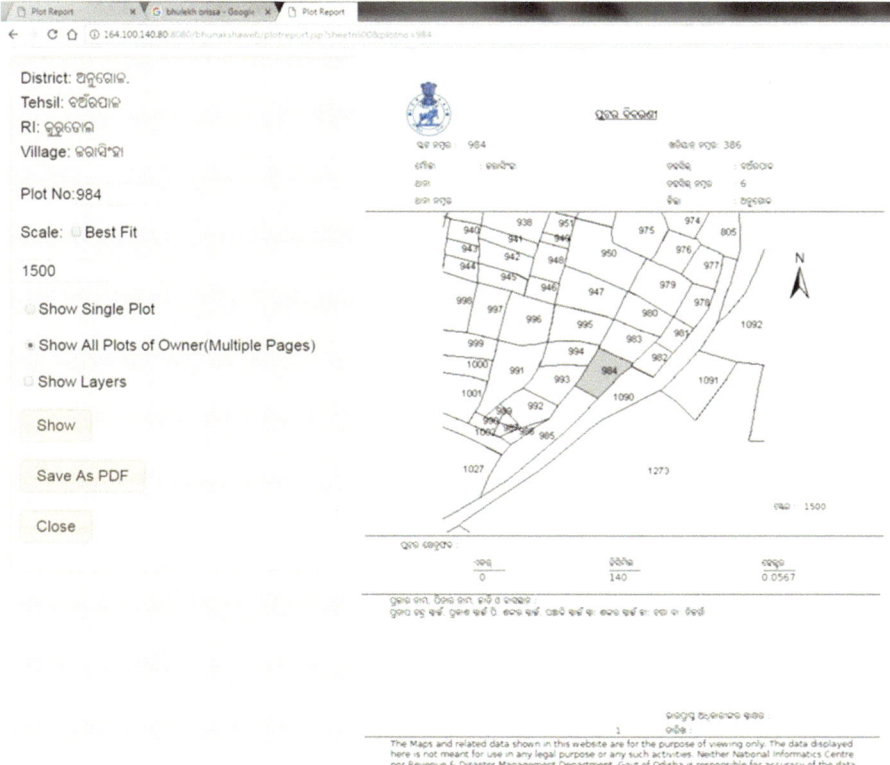

Picture 11.7 Official website (Odisha)providing the details of revenue information

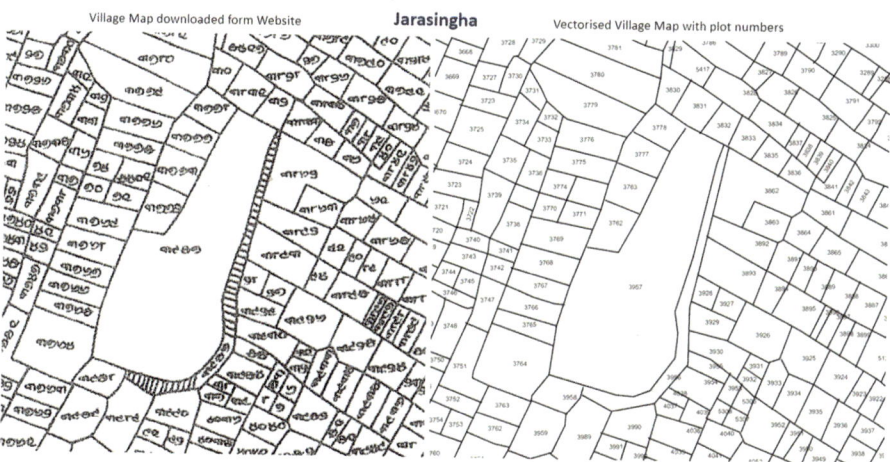

Picture 11.8 Village map of Jarasingha (partial) downloaded and vectorized for GIS-based inventory

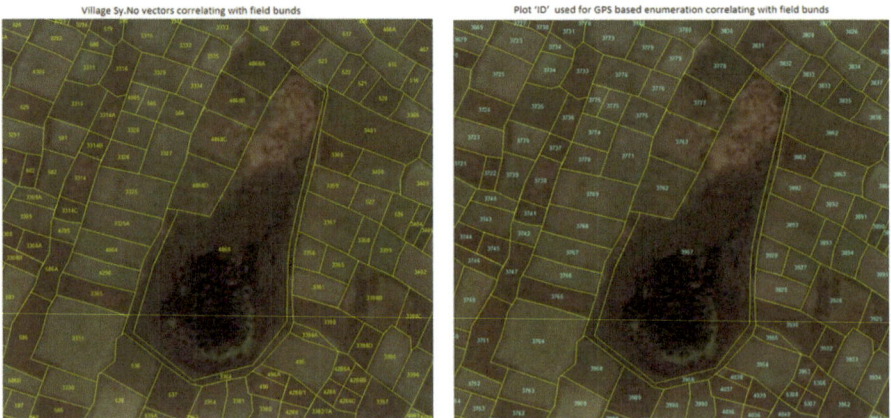

Picture 11.9 Correlation of plots from village map (Jarasingha) with field bunds (plots) extracted from image

	A	B	C	D	E	F	G	H	I	J
1	Khata_No	Farmer _Name	Plot_ID/Sy_no	GIS_Syno	Land_type	Description	Area_acre	Area_Gunta	Area_Decimal	Remarks
4716	780	Rakhita	375	375	ଚଷ ୧ଡ		0	70	0.0283	ଭ ୯ତ ଓା ନଚ ଓ ଠ ଓଠ
4717	780	Rakhita	376	376	ଚଷ ୧ଡ		0	50	0.0202	ଭ ୯ତ ଓା ନଚ ଓ ଠ ଓଠ
4718	780	Rakhita	377	377	ଚଷ ୧ଡ		0	120	0.0486	ଭ ୯ତ ଓା ନଚ ଓ ଠ ଓଠ
4719	780	Rakhita	3956	3956	ଷ ୯		0	330	0.1335	
4720	780	Rakhita	3957	3957	୨ ଣ ଭଣ ୧ଡ		2	680	1.0846	ଉଣ୍ ଣଦଠ
4721	780	Rakhita	4	4	ଚଷ ୧ଡ		0	20	0.0081	ଚ ଠ ଣ ଠ ଠଣ ଘ ଦାଟଣ୍ ୧ଚ ଠଟଣ, ଦ ଠ ା ଣ୍ ଚଚଠ ଠ ଚ, ଚଠ ଚ ଠ ଚ ଠ ଣ ଦ ଚ ଟ, 1970 ଠ ଣ ଠ ଠ ୯ଚ ଚ ଠଠ, ଚ ଠ
4722	780	Rakhita	42	42	ଠଠ୧ ଗ		0	290	0.1173	ନ ୯ ଠ ୯ ଣ ଠ ଠ, ୯ଚଠ ଠ
4723	780	Rakhita	4513/5150	4513/5150	ଠ ଠ		0	120	0.0486	ଠ ଠ୍ ଦଣ ଠ ଠ ଣ ନ ଦ ଠ
4724	780	Rakhita	4523	4523	ଚଷ ୧ଡ		3	120	1.2627	
4725	780	Rakhita	4525	4525	ଠ ଠ ଣ ଠ ୯ଠ		2	680	0.0846	ଷ ନ/28
4726	780	Rakhita	4538	4538	୧ ଠ ଣ ଠ ୯ଠ		0	240	0.0971	ଷ ନ/2
4727	780	Rakhita	4540	4540	ନ ୧ ଠ		0	20	0.0081	

Picture 11.10 Khata extracts of Jarasingha downloaded from the official website

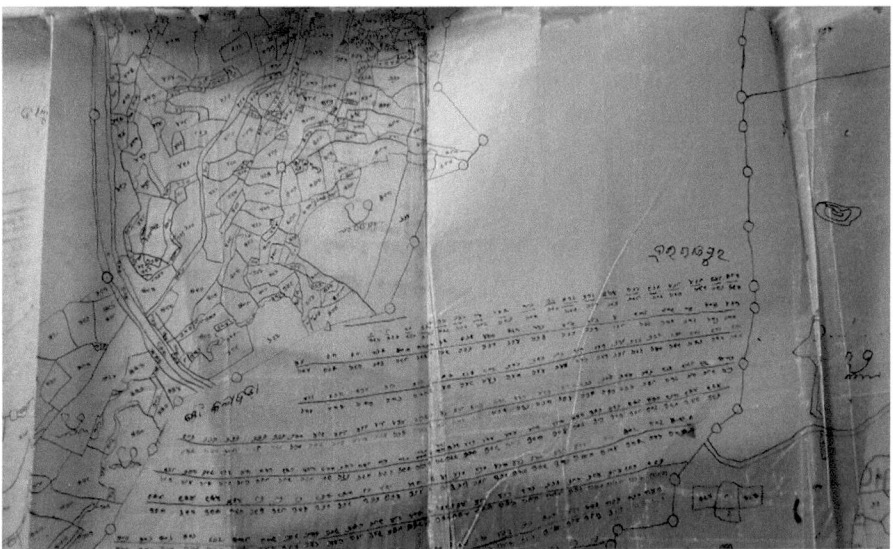

Picture 11.11   Map used by the officials for crop enumeration in Jarasingha

Chapter 12
Conclusion and Recommendations

Abstract ICRISAT conducted the research experiment with the hypothesis that agricultural statistics should correspond to ground situations and determine the actual extent of each crop along with cultivation and other farming practices. The process should provide reliable, retraceable geographical data that are climate independent and can be applied to different agroclimatic regions. Having studied different approaches, tools, and technologies for generating crop statistics, it was decided to adopt "geo-stamping" approach with different geospatial tools. It is an integration of remote sensing, GPS, and GIS having inbuilt tracking of workflow of any regular data collection and reporting system and allowing plot-level inventory.

Keywords Agricultural statistics · Retraceable geographic data · Geo-stamping · Workflow

12.1 Conclusions

Based on the experiment, a proper workflow of plot level crop enumeration, organizing the data, and analysis has been arrived (Fig. 12.1). Following are the conclusions that could be drawn from the experiments at different locations with different physiographic, climatic conditions and cultivation practices.

12.1.1 Generation of Crop Plot Data

High-resolution satellite images have been found to be useful in generating most reliable spatial reference systems/platforms. Cartosat images were explored and though found useful, considering the economical aspect, the images provided freely by Google Earth have been used. Extraction of crop plot boundaries has been a challenge. Different methods of digital extraction were tried, but manual method

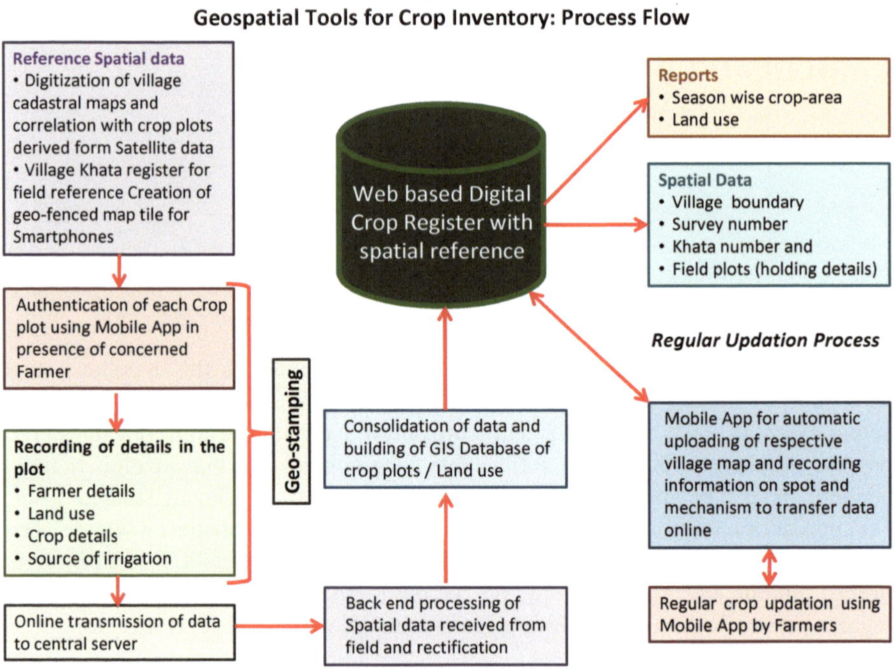

Geospatial Tools for Crop Inventory: Process Flow

Reference Spatial data
• Digitization of village cadastral maps and correlation with crop plots derived form Satellite data
• Village Khata register for field reference Creation of geo-fenced map tile for Smartphones

Authentication of each Crop plot using Mobile App in presence of concerned Farmer

Recording of details in the plot
• Farmer details
• Land use
• Crop details
• Source of irrigation

Online transmission of data to central server

Geo-stamping

Web based Digital Crop Register with spatial reference

Consolidation of data and building of GIS Database of crop plots / Land use

Back end processing of Spatial data received from field and rectification

Reports
• Season wise crop-area
• Land use

Spatial Data
• Village boundary
• Survey number
• Khata number and
• Field plots (holding details)

Regular Updation Process

Mobile App for automatic uploading of respective village map and recording information on spot and mechanism to transfer data online

Regular crop updation using Mobile App by Farmers

Fig. 12.1 Process flow for geospatial intervention for crop inventory

was found to be reliable and accurate. Visual interpretation of natural color composite was found to be useful in identifying and extracting the field bunds.

12.1.2   Registration of Village Revenue Maps

Mere crop acreage and type is not useful unless it is correlated with the farmer. Though revenue records of a cultivator have a record of crop, the geographic context and spatial reference to other holdings are missing. The experiment has been successful in registering the crop plots extracted from satellite images with village revenue survey maps. Further, based on the field inventory, each crop plot forming a holding within a given survey number was generated.

12.1.3   Geo-fenced Map Tiles for Automatic Navigation

Locating a parcel exactly in the field and correlating with a farmer are difficult processes. If the map is converted to geo-fenced tile for a specific village, one can conveniently navigate to the spot and record the site details. GIS technologies have been useful in converting the map with plot and survey number boundaries into a projected map tile to be able to conform with GPS system. Application developed on Android platform with GPS-enabled smartphones has been very useful in exactly identifying and registering coordinates of the plot and allowed to record the crop on the spot. In addition, few important aspects related to cultivation practice, method, sources of water, condition of crop, availability of inputs, and constraints of farming could be captured. The data captured on daily basis was downloaded and later integrated with GIS.

12.1.4   GIS for Visualization and Analysis

For effective analysis, any data needs to be properly compiled and integrated and organized. To bring in spatial context for the agricultural data, geographic coordinates of the crop plots along with shape and size were generated, and spot photograph of the crop was also taken. Entire dataset was systematically organized in QGIS so that one can get proper visualization of different crops and extent of each crop. Since the data was already captured in digital format at the source, no delay in further format editing and compilation was necessary. The multilayer integration facility provided in the GIS has been useful to analyze the interrelationship of different features, geographic distribution of crops, etc.

12.1.5   Inventory During Different Seasons

Agricultural statistics should be available for all the growing seasons and in this case, *Kharif*, *Rabi*, and *Summer* so that crops and cropping pattern in every season are registered and variations documented. Having registered all the crop plots during the first (in this case, *Rabi*, 2016–2017), the same GIS data was used for update. The Android software is modified in such a way to avoid geo-stamping of every plot again. If a plot has a crop, it is geo-stamped, lest enumerator can move to other plot. Avoiding redundant stamping of each plot saves time and consequently the cost. The experiment was successful in recording the only plot having crops and the rest being systematically documented in the field itself. The time saved during the second time of inventory is almost one third and provided indication of its being able to be scaled up.

12.1.6   Inventory Independent of Weather Condition

Since a GPS-based mobile device is being used in the field for crop inventory, rainfall, cloud, etc. do not affect the survey. One can conduct the inventory during any season. The application and the reference database can be used to monitor the crop health as well during the growing season. In case of any effects like flooding, pest attack, etc., the application developed could be used to authenticate the actual area by geo-stamping the affected plots.

12.1.7   Online GIS and Real-Time Data Synchronization

Though real-time data was captured using GPS in the field, data organization had a gap of couple of days since data synchronization was not automatic. This was a cautious step to avoid loss of time in the field due to uncertainty of bandwidth and/ or Internet strength in the remote places of the villages. The Android application was modified to transmit the data to central GIS server during *Rabi* season in Khatijapura. The back-end GIS application was scaled up to web-based system, and data structure was modified. The field tests were successful, and data was synchronized with server directly from the field. At the back end, the plots for which data was captured could be neatly observed, and there was no preprocessing of the data. One could directly visualize the changes in the datasets of two different seasons.

12.1.8   Standard Process for Crop Inventory

Based on the action research and with continuous refinements from season to season, a workable model for crop inventory has been developed (Fig. 12.1). The process conforms to the standard requirements of crop statistics, and the field trials with local youths have also been successful indicating that data generation process could be a participatory process as well.

12.2   Recommendations

Intervention of geospatial technology for crop inventory in particular and agriculture in general has the following advantages:

- Enables visualization of crop area by type and changing trends
- Enables gaining critical insight without additional effort
- Has inbuilt workflow of data collection and reporting system
- It helps to achieve:
  - Better monitoring of efficiency of process
  - Better and effective recording of crops
  - Better reporting of location-specific details and eventually predict outcomes

The potential uses of the data are as follows:

(a) Small and marginal and also large farmers
(b) Water user associations
(c) Farmer producers organization
(d) Agricultural entrepreneurs
(e) Government departments including agricultural finance and insurance agencies
(f) Private agencies

The data generated on time helps to improve efficient use of:

- Agricultural inputs
- Outputs
- Crop diversification
- Access to ground and surface water
- Enabling drip and sprinkler irrigation
- Crop productivity
- Water use efficiency
- Institutional strengthening
- Capital investment by government and private agencies
- Easy and efficient digital communication, dissemination, and advisories to all stakeholders
- Farm mechanization
- Postharvest, primary processing, and agribusiness promotion
- All decision-makers get online- and mobile-based information

The geospatial interventions for generating agricultural data should be scaled up. The process may be taken up in phases to cover the areas wherein agriculture is dependent on rainfall alone and gradually to other regions.

12.3 Main Outputs of the Study

The research work in three villages spread across different states has resulted in the following important outputs:

- Accurate spatial information related to crops, crops of different seasons, and their geographical variation during different seasons.
- Software application integrated with GIS which allows generating different MIS reports.
- Spatial database of individual holdings in GIS environment facilitating search of any holding in the village allowing traceability and reusability.
- GPS software for crop mapping in different seasons.
- Model for quick compilation of field data.

Chapter 13
SWOT Analysis of Geospatial Technology

Abstract Geospatial tools are useful for generating crop statistics. Authenticity, error-free data compilation, GIS-based area computation, traceability, and estimation of area of mixed crops are the strengths of the technologies. Updation of changing field bunds through GPS, possibility of non-availability farmers, and "not for legal purpose" appear to be the weaknesses of the method advocated. Possibility of consolidation of holding level data, upgradation of skills of personnel, and need of the data in various public and private sectors are the opportunities, and the reluctance to adopt the technology and the financial allocation from the government may pose threat to useful geospatial technology adoption for agriculture statistics.

Keywords Error-free data · GIS-based area computation · Changing field bunds · Consolidation of land records · Skill upgradation · Reluctance to adopt · Financial allocation

Since the current research experiment is oriented toward evaluating the usefulness of the geospatial technological interventions to the existing system for generation of crop area statistics, a SWOT analysis was conducted.

13.1 Strengths

1. *Authenticity of the information:* Field enumeration details are recorded on the spot showing the geographic coordinates of the crop plot and time of inventory along with photograph. In case of any variations within a plot, options are available to trace and change which are automatically captured.
2. *No mistakes in data entry and compilation:* Data is recorded on the spot digitally and gets automatically integrated with the master table. Any further report is derived from the master database, and hence data entry errors like oversight, mistakes, etc. are completely avoided.

3. *Area calculation is based on GIS:* Every crop plot is already organized in GIS extracted from the satellite imagery, and crop type is recorded on the spot, completely avoiding the recording of crop area information based on previously available area mentioned in the RTC.
4. *Traceability:* Since the data is based on coordinates and systematic identification of each plots, errors, if any, can be accurately and easily traced giving no scope for manipulation.
5. *Accurate calculation of area under mixed crops:* Since the field enumerator registers the cropping pattern in the field related to mixed crops, the standard formula adopted by department can be built into the software application; area under two different crops in a single plot is calculated automatically.
6. *Timely information:* Within very short time of field inventory (within a day normally, if there are spatial changes maximum of a week), the crop area statistics are available, which helps in estimating the production (based on the standards from crop cutting experiments).
7. *Determination of crop loss due to various reasons:* Since the information is already in GIS, mobiles can be used to identify affected areas (flood, diseases, etc.).
8. *Outsourcing:* Once the master database of any district/taluk is developed and integrated with application, there is a possibility of outsourcing the subsequent process of crop enumeration. Separate agencies can be fixed for a specific district based on competitive tender for GPS-based enumeration for any cropping season. The selected agency provided the district−/taluk-specific data and GPS application for enumeration. There will not be any capital cost on the GPS equipment as the agency would carry its equipment. Since reliability and accuracy is bound with GPS, there would not be any issues on these aspects. This could be a regular process. Further, it will not create a concern even if technology gets refreshed.

13.2   Weakness

1. *Spatial changes have to be traced with GPS:* In case of any crop plots have undergone spatial changes (if divided into two plots or within the plot two crops are cultivated), it is required to trace the same with GPS. *Though it adds to time of enumeration depending upon the extent of changes, it reliably updates the changes.*
2. *Possibility of missing changes of ownership:* Enumerator needs not consult the farmer on routine inventory (during the second round). If there are any changes of ownership due to sale or other such reasons and if the farmer is not available for consultation or if it is not updated in the statutory records, enumerator will not be able to record the ownership details. However, the crop area figures will not have any variation.

3. *Cannot be used for legal purpose:* Though methodology enables generation of holding map, it is only for visualization of area and used for crop area statistics and hence cannot be used for any other legal purposes. The holding maps are not ratified through the revenue administration.

13.3   Opportunity

1. *Update of spatial data:* So far no integrated spatial data related to sub-survey numbers and holdings pattern are available either in print or digital format. This intervention provides best opportunity for developing comprehensive digital spatial information related to land holding pattern. This would be of great inputs to various other purposes.
2. *Opportunity to upgrade the skill with officials:* The officers involved in the process will have an opportunity to upgrade their skills in visualization, analysis, and generation; thereby skill upgradation happens, and efficiency and effectiveness of process improve.
3. *Opportunity to update the information in RTC:* With the information obtained from the field being current, reliable, and accurate, the same can be entered into respective RTCs for update as regards crop details concerned.
4. *Opportunity to sell the data:* Reliable information on crop area are required by banks and insurance companies. This provides a good opportunity for government to provide access to these financial institutions at cost.

13.4   Threats

As such there are no threats for the technology intervention. However, the following could be listed as possible threats:

13.4.1   Internal

1. *Officers' reluctance to upgrade skills:* Officers involved in the grassroot level may not show interest to upgrade themselves with interventions or may not use the gadgets for the purpose.
2. *Budgetary allocation:* Though the intervention brings efficiency and effectiveness and accuracy to the information system, initial costs may have to be justified. However, this can be taken up in phased manner and thereby cover entire state.

13.4.2   External

1. *Farmers may not cooperate:* Reluctance to show the holdings and provide correct information has been observed during the field survey. Education and proper campaigns would provide better results. However, if local youths are involved in the process, it might demystify the concerns.

Chapter 14
Geospatial Technology Intervention: A Best Method for Crop Inventory

Abstract Challenge of area estimation well within time growing season and documenting the area of mixed crops and measuring the extent of seasonally varying crops can be conveniently and efficiently achieved by geospatial technology. As the technology is made simple, adaptability of the technology by rural youths provides opportunity for temporal financial support and thereby scalable. Cost of generating all the related data to agricultural holdings by means of geospatial technology is less than 1% for an annual household earning. Compared to cost of gathering the agricultural statistics by employing the village accountants and the governmental overheads, the cost of data generation using geospatial tools is economical.

Keywords Area estimation · Mixed crops · Adaptability · Scalability · Earning of house hold · Geospatial tools

Crop production estimates generally include two aspects: (1) acreage/area covered by crop/crops and (2) yield (expected/determined) per unit area. Though more focus is on crop yield, there are many complexities to the estimation of area. The challenge is to estimate the area well within the growing season that ultimately provides inputs to production estimation. Heterogeneous cropping pattern in a single season and different crops in different seasons introduces more complexity. There have been various approaches and practices of crop area estimation. Statistical methods involving different kinds of sampling have been popular, but they do not always present situations on the ground.

Accuracy, authenticity, and reliability are the important issues when it comes to crucial information that goes into planning. There are useful methods that generate information at national and state level. When it comes to village level, as the worlds' goal is toward doubling the income of farmers and improved agricultural productivity and climate resilience, there is a need for spatialized information about agricultural practices. Geospatial technologies appear to be the best-suited interventions. Since these technological tools are made in the process of workflow and geographical context, it helps a lot during exigencies like flood and drought situations for aptly identifying the affected areas. These tools have the potential to replace transact or grid sampling.

The experiments in three locations have clearly exhibited the advantage and the accuracy and, more than everything, quick compilation of the data and providing the results. The tools can also be adopted anytime during the growing seasons so that the area affected by pests and any other calamities could be properly estimated and crop area that could be ultimately available for harvest is estimated in advance.

It needs to address the effectiveness of the solution/intervention. The experiment has demonstrated that if once the base information along with ownership details are generated, the efforts of gathering the data in any growing season are only systematic field enumeration which does not require experts. All the formats and procedures are simple and could be understood by any undergraduate. Since mobile phones have become the common means of communication and almost every household in any village uses it, technology adoption is not a constraint.

The experiment revealed that the cost for generating information using the geospatial tools is about Rs.700/hectare for 1 complete agriculture year (three seasons) which includes the travel, food, and manpower and also the technology infrastructure. This cost is for the first year wherein the spatial database of all the holdings is prepared. Second year onward, it is only the cost for field visits using the gadgets (smartphones) to register the crop. It would reduce to less than Rs.400/hectare per year.

Farm households earned INR 77,888 in the period from July 2012 to June 2013 or INR 6491 per month during this period. Across landholding classes, the lowest land class (with less than 0.01 ha land) earned INR 54, 147 in the period, while the largest land class (with greater than 10 ha land) earned INR 4, 52,299 in the period (Ranganathan 2015). The national goal is toward doubling the income of farmers and improved agricultural productivity. Understanding critical issues at holding level with respect to cultivation practices and constraints of the farmer is crucial for achieving this goal. Cost of generating all the related data to agricultural holdings by means of geospatial technology is less than 1% for 1 year of earning of one household. This cost is for one-time generation of exhaustive information. The recurring cost keeps on reducing and has a chance to become almost nil if the farmers are enabled or few farming youths are involved in the process. Alternatively, once the basic data is generated, every village having a moderate size of agricultural lands of about 500 ha. has a prospect of providing part-time employment generating an income of at least INR 30000 per year. The resources could be from one family or multiple youths who can share the enumeration work, and it adds that much income for the participating families.

Compared to cost of gathering the agricultural statistics by employing the village accountants and the governmental overheads, the cost of data generation using geospatial tools is economical. It has a great opportunity of creating partial employment in the rural area for the youths who could take up the exercise once during the growing season. Since each crop plot is registered and enumerated and the data gets compiled very fast, there is no time lag in disseminating the results. One can get the acreage information of every crop grown in every season.

Chapter 15
Executive Summary

Abstract Ending poverty and hunger by 2030 being one of the core objectives of Sustainable Development Goals (World Bank 2015) speaks of doubling the income and the need for improved agricultural productivity and climate resilience, strengthened links to markets, agribusiness growth, and rural nonfarm incomes. Reliable and real time information related to agriculture is essential to address the agriculture growth in developing nations. Agricultural statistics including holding by distribution, size, tenure, land use, means of production are increasingly becoming important. The quality of available agricultural data and the methods by which such data is collected are weak in several developing countries. Geo-stamping approach i.e., the Remote Sensing (RS), Geographic information System (GIS) and Global Positioning System (GPS) intervention adopted in conformation with existing work flow for crop-area enumeration has yielded reliable results.

Ending poverty and hunger by 2030 being one of the core objectives of Sustainable Development Goals (World Bank 2015) speaks of doubling the income and the need for improved agricultural productivity and climate resilience, strengthened links to markets, agribusiness growth, and rural nonfarm incomes. Reliable and real time information related to agriculture is essential to address the agriculture growth in developing nations. Agricultural statistics are part of the economic profile of a village/Taluk/District /State and country and are increasingly becoming important.

Agricultural data includes agricultural holding by distribution, size, tenure, land use, means of production, and labor force. In addition to its inevitable role in food security, agricultural development is now seen as a vital and high-impact source of poverty reduction. The links between poverty and crop yields, depending upon a variety of factors like cultivation practices, availability of irrigation, and access to resources to buy agricultural inputs for adoption of new technology, cannot be fully understood without reliable information of crop area and types. In the absence of reliable information on crop productivity, the reasons behind food insecurity of agricultural households cannot be precisely identified.

Though importance of agriculture in economic development and food security is being addressed, the quality of available agricultural data and the methods by which such data is collected are weak in several developing countries. As of now crop area estimates have problems associated with both accuracy and time availability. Crop area data are generally available a few months after harvest. Having reliable data before harvest is a major challenge. There are useful methods that generate information at national and state level. When it comes to village level, there is a need for spatialized information about agricultural practices.

Review of different approaches adopted in various parts of the world for generating the crop area statistics indicates that crop area information is not available well before the harvest. The method adopted by the concerned departments in the country has constraints as follows:

- Enumerators do not visit the field during the specified time for inventory.
- Enumerators tend to use the old information (previously available) and pass for further compilation.
- Data compilation happens at different levels, and it takes time for aggregation. The errors, if any; go unnoticed.
- The departments do not deploy adequate human resources for the work (understaffed as revealed during the discussion with departments).

In order to address the gap of reliable agricultural statistics at the village level, ICRISAT conducted action research in three different agroclimatic regions of India by using geospatial tools. The theme of the current research has been exploring means of generating reliable information on crops during the growing season. In this context, the following hypotheses have been postulated, and experiments were conducted:

- The statistical data on crops derived by means of extracting the old information from the revenue records does not match with ground situations.
- Current pattern of crop statistics does not provide information on all types of crops grown in any season.
- Geographical context of crop statistics provides microlevel information related to crops, and holding level information is essential to understand the constraints of cultivation and other farming practices.

The research experiments were executed in different agroclimatic regions of the country. As a part of the study, Khatijapura (Vijayapura District of Karnataka), Daliparru (Krishna District of Andhra Pradesh), and Jarasingha (Angul District of Odisha) were selected in consultation with respective state agencies.

Geo-stamping approach, i.e., the remote sensing (RS), geographic information system (GIS), and global positioning system (GPS) intervention, has been adopted in conformation with existing workflow for crop area enumeration. The process involved:

- Generating DGPS control points for image rectification, extraction of crop plot boundaries from the existing high-resolution satellite images in a GIS environment.

- Development of a geo-referenced controlled mosaic of crop plots with revenue survey numbers (village revenue survey maps).
- Creating geo-referenced map tile for loading on to a GPS-based smartphone.
- Development of geo-referenced map tile, suitable android application for GPS-based smartphone for automatic navigation to the crop plot and recording the field details on the spot.
- Creation of GIS database of the villages with crop information visualization.

Experiment was conducted simultaneously in all the three villages during *Rabi*, *Summer* (2016–2017), and *Kharif* (2017–2018).

Plot-level crop inventory was carried out using GPS-based mobile devices in all the three villages. Information related to few socioeconomic aspects have also been collected from the farmers. Field records from the GPS-based smartphones were downloaded on a daily basis. The data was integrated with GIS database of respective villages that have the survey number-wise information.

Brief results of the experiments are as follows:

1. Khatijapura, a rainfed semiarid dryland region, has cultivation spread out mainly during *Kharif* and *Rabi* seasons. Area cultivated during *Rabi* is more compared to *Kharif*, and during *Summer*, lands are left vacant, and very limited crops are grown with groundwater. During *Kharif*, tur is the main crop with few pulses (black gram) and vegetables. Total area cultivated during *Rabi* is 293.01 ha, while it is 151.17 ha. in *Kharif*.
2. Daliparru, a coastal region of Andhra Pradesh, has irrigation facility, and the entire village is dominated (almost 99.5%) by paddy during *Kharif* and by black gram cultivation during *Rabi* with very little extent of other crops.
3. Jarasingha in Angul District of Odisha presents a different scenario. During *Kharif*, the maximum area is under paddy as there is a major irrigation facility. It is also characterized by the presence of substantial area under vegetable crops.
4. Jarasingha village has a unique feature of kitchen garden wherein people grow vegetables of different varieties that yield substantial livelihood option. There are 291 plots that have different crops, and cumulative area during September 2017 was 6.03 ha. About 203 households have kitchen garden.
5. Cultivation practices vary from village to village. In Khatijapura almost 30% of the landowners reside in nearby Vijayapura city and practice contract farming (17%), crop sharing (6%), and leasing out (1%) of farmlands. Daliparru is located about 18 km from Machilipatnam. About 15% of land belongs to migrants and 7% to NRIs. Jarasingha village has two settlements – New Jarasingha on SH-63 and old Jarasingha nearby river. Since the village is located 5 km from Angul city, almost all the big landowners reside in Angul city and practice contract farming, crop sharing, and leasing out of farmlands.

Comparison of the results with the data of from respective state departments indicates the following:

- There are differences not only in the number of crops recorded but also in the actual area covered by each crop.

- The crop abstract records in case of Khatijapura do not reflect all the crops grown during *Rabi* 2016–2017. The net area sown is almost one third when compared to the actual as enumerated by GPS. Also the land utilization data is not updated.
- The available data in case of Daliparru is almost near to the one generated from GPS intervention. However, the records do not mention the minor crops grown during the season indicating that enumeration is not done on plot-to-plot level.
- In Jarasingha, the winter season corresponding to *Rabi* shows certain difference in the data. GPS inventory indicates no paddy, while official records show 125.6 ha, and acreage of other crops do not correlate. The acreage of other vegetables also has substantial variation (GPS, 47.74 ha; and department, 10.95 ha).
- GPS inventory is able to generate reliable and plot-level (farmer level/holding level) data, whereas the existing method fails to produce this information.
- Revenue records: In all the states, the crop statistics are based on revenue records. All the states, though have good revenue records in the form of maps and attribute (*Khata* extracts/information), appear to have not used them for crop enumeration. The discrete sketches have not been properly mosaicked, and no composite/integrated holding (*Khata*) maps are available. In Jarasingha, the maps being referred to are totally different than what is available on the official website.
- GPS-GIS intervention has been able to generate a systematic and useful land holding maps that clearly show the disposition of lands of different holdings.

The following conclusions could be drawn from the research experiment:

- High-resolution satellite imagery can be used for extracting the crop plot boundaries which in turn correlate well with the individual holdings. GPS intervention is useful to record the details of crop and that of farmer along with practices on the ground on the spot.
- The use of android app on smartphone with geo-referenced map of the village is useful to navigate to any field and record the details of crop on field itself. The method for the first-time inventory provides facility of registration once for all. Enumeration time can be reduced with slight modification of the application, and field traverse can be optimized.
- With the help of android application loaded on smartphone, any small changes in the boundary of a plot can be captured and updated in the GIS.

The research work in three villages spread across different states has resulted in the following important outputs:

- Accurate spatial information related to crops, crops of different seasons and their geographical variation during different seasons.
- Software application integrated with GIS which allows generation of different MIS reports.
- Spatial database of individual holdings in GIS environment facilitates to search any holding in the village allowing traceability and reusability.
- GPS software for crop mapping in a different season.

- A means of eliminating errors at the time of data entry, compilation, and computerization.
- Quick compilation of field data.

Based on the experience of conducting the pilot at three villages in different agroclimatic regions and landscapes and results obtained, the following recommendations are made:

- The interventions should be scaled up as it improves the crop statistics in terms of accuracy, authenticity/reliability, traceability, timeliness, access, and analysis. Since the data is organized as per the administrative hierarchy, crop area information, the holding level is available, and there is no delay in compilation.
- Any person who can use mobile will be able to carry out crop enumeration using this method, and training is simple. If educated youth in the villages are deployed for repeated enumeration, it provides temporary employment opportunity.

High-resolution satellite data (latest) should be used to extract plot boundaries so that time for generating basic spatial data is reduced.

Glossary

AA Agricultural Assistant

AAO Assistant Agricultural Officer

AE Advance Estimates

AO Agricultural Officer

ASO Assistant Statistical Officer

CITARS Crop Identification Technology Assessment for Remote Sensing

CPO Chief Planning Officer

DAC & FW Department of Agriculture Cooperation & Farmers Welfare

DCS District Crop Abstract

DES Department of Economics and Statistics

DGPS Differential Global Positioning Systems

DPMU District-Level Planning and Monitoring Units

DRO District Revenue Officer

DSO District Statistical Officer

EARAS Establishment of an Agency for Reporting Agricultural Statistics

FASAL Forecasting Agricultural output using Space, Agrometeorology, and Land-based observations

GAGAN GPS-aided GEO Augmented Navigation

GCP Ground Control Point

GIS Geographic Information System

GNSS Global Navigation Satellite System

GPS Global Positioning System

Ha Hectare

HCS Hobli Wise Report

ICRISAT International Crops Research Institute for the Semi-Arid Tropics

ICS Improvement of Crop Statistics Scheme

ICT Information and Communication Technology

ISRO Indian Space Research Organization

IIM Indian Institute of Management

IMD India Meteorological Department
LACIE Large Area Crop Inventory Experiment
LIDAR Light Detection and Ranging
LISS Linear Imaging Self-scanning Sensor
MIS Management Information System
MNCFC Mahalanobis National Crop Forecast Centre
MRO Mandal Revenue Officer
NRI Nonresident Indian
NRSC National Remote Sensing Center
NSSO National Sample Survey Organisation
ORSAC Odisha Space Application Centre
PCA Pilot Census of Agriculture
PIXEL Pixel Softek Pvt. Ltd.
QGIS Quantum Geographic Information System
RDO Revenue Divisional Officer
RI Revenue Inspector
RS Remote Sensing
RTC Records of Rights, Tenancy and Crops
SAC Space Applications Center
SASA State Agricultural Statistics Authorities
SBAS Satellite-Based Augmentation System
SFI Statistical Field Inspectors
SFS Statistical Field Surveyors
SO Statistical Officer
SOP Standard Operating Procedure
SPP Single Point Positioning
Sy. No. Survey Number
TCS Taluk Crop Abstract
TRS Timely Reporting Scheme
U.S. United States of America
UT Union Territory
UTM Universal Transverse Mercator
VA Village Accountant
VAW Village Agricultural Worker
VCS Village Crop Statistics
VHR Very High Resolution
VPCS Village Panchayat Wise Report
VRA Village Revenue Assistants
VRO Village Revenue Officer
WGS84 World Geodetic System 1984
WI Works Inspector
ZOOMIN Zoomin Softech Pvt. Ltd.

References

Alemu MM (2016) Automated farm field delineation and crop detection from satellite images: thesis from geo-information science and earth observation faculty of University of Twente, The Netherlands

Anonymous (2005) The guidelines issued by the Government for generating crop area statistics (Government Notification No. RD 23 ELR 2004 dt.06.05.2005)

Bailey J, Boryan C (2010) Remote sensing applications in agriculture at the USDA National Agricultural Statistics Service. Research and Development Division, USDA, NASS, Fairfax, VA

Bégué A, Arvor D, Bellon B, Betbeder J, de Abelleyra D, Ferraz RPD, Lebourgeois V, Lelong C, Simões M, Verón SR (2018) Remote sensing and cropping practices: a review. Remote Sens 10:99. https://doi.org/10.3390/rs10010099

Breidt F, Fuller W (1999) Design of supplemented panel surveys with application to the National Resources Inventory. J Agric Biol Environ Stat 4(4):391–403

Carfagna E, Keita N (2009) Use of modern geo-positioning devices in agricultural censuses and surveys. Bulletin of the International Statistical Institute, the 57th Session, proceedings, special topics contributed paper meetings (STCPM22) organized by Naman Keita (FAO), Durban, August 16–22, 2009

Central Statistical Organisation (2008) Manual on area and crop production statistics

Cotter J, Davies C, Nealon J, Roberts R (2010) Chapter 11. Area frame design for agricultural surveys. In: Benedetti R, Bee M, Espa G, Piersimoni F (eds) Agricultural survey methods. Wiley, Chichester

Craig M, Atkinson D (2013) A literature review of crop area estimation: for UN-FAO

Crnojevic V, Lugonja P, Branko NB, Brunet B (2014) Classification of small agricultural fields using combined Landsat-8 and rapid eye imagery – case study of Northern Serbia. J Appl Remote Sens 8(1):083512

Dadhwal VK, Singh RP, Dutta S, Parihar JS (2002) Remote sensing based crop inventory: a review of Indian experience. Trop Ecol 43(1):107–122

Davies C (2009) Area frame design for agricultural surveys. RDD research report, research and development division, USDA-NASS, Fairfax, VA

FAO (2005) The state of food and agriculture 2005. FAO, Rome ISBN 92-5-105349-9

Ferreira SL, Newby T, du Preez E (2006) Use of remote sensing in support of crop area estimates in South Africa. Compilation of ISPRS WG VIII/10 Workshop 2006, remote sensing support to crop yield forecast and area estimates, EC JRC, Stresa, Italy

Gallego FJ (1995) Sampling frames of square segments. MARS project, Institute for Remote Sensing Applications, JRC, Report EUR 16317. ISBN 92-827-5106-6

Gallego FJ (1999) Crop area estimation in the MARS Project. Agriculture and Regional Information Systems, Space Applications Institute, JRC

Gallego FJ (2006) Review of the main remote sensing methods for crop area estimates agriculture unit. Compilation of ISPRS WG VIII/10 Workshop 2006, remote sensing support to crop yield forecast and area estimates, Stresa, Italy, Agriculture Unit, IPSC, JRC

Gallego FJ (2011) The use of area frame surveys and remote sensing. Agriculture unit, IPSC, JRC

Gallego J, Craig M, Michaelsen J, Bossyns B, Fritz S (2008) Workshop on best practices for crop area estimation with remote sensing data: summary of country inputs. Group on earth observations (GEO), GEOSS Community of Practice Ag Task 0703a, EC JRC, Ispra, Italy

Hearst Anthony A (2014) Automatic extraction of plots from geo-registered UAS imagery of crop fields with complex planting schemes. Open access thesis paper 332. Purdue University

Hegde VR, Jayaraj KR, Karale RL, Subba Rao P (1994) Area estimation of arecanut plantation in Sirsi Taluk using IRS data. J Ind Soc Remote Sens 22(3):149

Hooda RS, Yadav M, Kalubarme MH (2006) Wheat production estimation using remote sensing data: an Indian experience. Compilation of ISPRS WG VIII/10 Workshop 2006, remote sensing support to crop yield forecast and area estimates, Stresa, Italy, Agriculture Unit, IPSC, JRC

Huddleston HF (1978) Sampling techniques for measuring and forecasting crop yields. Economics, statistics and cooperative service (now NASS), USDA, Washington, DC

IIMB and ZOOMIN (2013) Crop inventory & updation technological intervention for increasing accuracy and value addition to current system. Research Report submitted to Karnataka Statistical System Development Agency (KSSDA) June 2013

Jain M, Mondal P, DeFries RS, Small C, Galford GL (2013) Mapping cropping intensity of smallholder farms: a comparison of methods using multiple sensors. Remote Sens Environ 134: 210–223

Jinguji I (2014) Dot sampling method for area estimation: crop monitoring for improved food security. In: Srivastava MK (ed) Proceedings of the expert meeting Vientiane, Lao People's Democratic Republic, 17 February 2014. Food and Agriculture Organization of the United Nations and the Asian Development Bank, Bangkok

Kussul N, Shelestov A, Skakun S, Kravchenko O, Moloshnii B (2012) Crop state and area estimation in ukraine based on remote and in situ observations. EC JRC, Ispra, Italy

Labuguen R, Anna Christine N, Durante D, Rotairo LE (2014) Adoption of Agricultural Land Information System (ALIS) for agricultural area estimation. Expert meeting on crop monitoring for improved food security 17 February 2014, Vientiane, Lao PDR

Lochan R (2006) System of collection of agricultural statistics in india including land use and area statistics. Internal report, Directorate of Economics and Statistics, New Delhi, India

Lu D, Batistella M, Li G, Moran E, Hetrick S, Freitas C, Dutra LV, Sant'Anna SJS (2012) Land use/cover classification in the Brazilian Amazon using satellite images. Pesq Agropec Bras 47(9):1185–1208. https://doi.org/10.1590/S0100-204X2012000900004

MacDonald RB (1984) A summary of the history of the development of automated remote sensing for agricultural applications. IEEE Trans Geosci Remote 22:473–481

MacDonald RB, Hall FG (1980) Global crop forecasting. Science 208(4445):670–679

Mike C (2010) A history of the cropland data layer at NASS. Research and Development Division, USDA, NASS, Fairfax, VA (Unpublished manuscript)

Miller H (2010) The data avalanche is here: shouldn't we be digging? J Reg Sci 50(1):181–201

Naik G, Basavaraj KP, Hegde VR, Paidi V, Subramanian A (2013) Using geospatial technology to strengthen data systems in developing countries: the case of agricultural statistics in India. Appl Geogr 43:99–112

Narciso G, Baruth B, Klisch A (2008) Crop area estimates with Radarsat: feasibility study in the Toscana Region – Italy. Internal report, JRC, IPSC – Agriculture Unit

Nusser SM, Goebel JJ (1997) The National Resources Inventory: a long-term multi-resource monitoring programme. Environ Ecol Stat 4(3):181–204

Ozdarici A, Ok AO, Schindler K (2015) Mapping of agricultural crops from single high-resolution multispectral images—data-driven smoothing vs. parcel-based smoothing. Remote Sens 7:5611–5638. https://doi.org/10.3390/rs70505611

Özüm Durgun Y, Gobin A, Van De Kerchove R, Tychon B (2016) Crop area mapping using 100-m PROBA-V time series. Remote Sens 8:585. https://doi.org/10.3390/rs8070585

Panda SS, Hoogenboom G, Paz JO (2010) Remote sensing and geospatial technological applications for site-specific management of fruit and nut crops: a review. Remote Sens 2:1973–1997. https://doi.org/10.3390/rs2081973

Patil AK (2012) Kautilya's views on agriculture. https://www.researchgate.net/publication/232 711210

Pender J (2007) Agricultural technology choices for poor farmers in less-favored areas of South and East Asia: IFPRI discussion paper 00709

Petja B, Nesamvuni E, Nkoana A (2014) Using geospatial information technology for rural agricultural development planning in the Nebo Plateau. S Afr: J Agric Sci 6(4):2014

Rakwatin P, Prakobya A, Sritarapipat T, Khobkhun B, Pannangpetch K, Sobue S, Oyoshi K, Okumura T, Tomiyama N (2014) Rice crop monitoring in Thailand using field server and satellite remote sensing: expert meeting on crop monitoring for improved food security 17 February 2014, Vientiane, Lao PDR

Ranganathan T (2015) Farmers' income in India: evidence from secondary data. Agricultural Economics Research Unit (AERU) Institute of Economic Growth (IEG), New Delhi

Raskar V (2013) IGDSS - applied technology at the grassroots using satellite imagery and mobile GPS at village level for onsite plot level crop-mapping. Int J Sci Res Publ 3(5):2250

Schøning P, Apuuli JBM, Menyha E, Muwanga-Zake ESK (2005) Handheld GPS equipment for agricultural statistics surveys experiments on area-measurement and geo-referencing of holdings done during fieldwork for the Uganda Pilot Census of Agriculture, 2003: Statistics Norway and Uganda Bureau of Statistics, November 2005

Singh RP, Sridhar VN, Dadhwal VK, Jaishankar R, Neelakanthan M, Srivastava AK, Bairagi GD, Sharma NK, Raza SA, Sharma R, Yadav M, Joshi FK, Purohit NL (2005) Village level crop inventory using remote sensing and field survey data. J Ind Soc Remote Sens 33(1):93–98

Song Q, Qiong H, Zhou Q, Hovis C, Xiang M, Tang H, Wenbin W (2017) In-season crop mapping with GF-1/WFV data by combining object-based image analysis and random forest. Remote Sens 9:1184. https://doi.org/10.3390/rs9111184

State of Indian Agriculture (2015-16) Government of India Ministry of Agriculture & Farmers Welfare Department of Agriculture, Cooperation & Farmers Welfare Directorate of Economics & Statistics New Delhi

Townsend R (2015) Ending poverty and hunger by 2030: an agenda for the global food system. World Bank Group, Washington, DC

Vijay S, Joy J, Apurva D, Kalubarme Manik H (2013) Geo-informatics and remote sensing applications for village level crop inventory in Gujarat state, India. Asian J Geoinf 13(2):2013

World Bank Report (2015) Report No. 95768, ending poverty and hunger by 2030: an agenda for the global food system

Index

© The Author(s), under exclusive licence to Springer Nature Switzerland AG 2019   149
K. V. Raju et al., *Geospatial Technologies for Agriculture*, SpringerBriefs
in Environmental Science, https://doi.org/10.1007/978-3-319-96646-5